Disruptive Tourism and its Untidy Guests

Leisure Studies in a Global Era
Series Editors:
Karl Spracklen, Professor of Leisure Studies, Leeds Metropolitan University, UK
Karen Fox, Professor of Leisure Studies, University of Alberta, Canada

In this book series, we defend leisure as a meaningful, theoretical, framing concept; and critical studies of leisure as a worthwhile intellectual and pedagogical activity. This is what makes this book series distinctive: we want to enhance the discipline of leisure studies and open it up to a richer range of ideas; and, conversely, we want sociology, cultural geographies and other social sciences and humanities to open up to engaging with critical and rigorous arguments from leisure studies. Getting beyond concerns about the grand project of leisure, we will use the series to demonstrate that leisure theory is central to understanding wider debates about identity, postmodernity and globalisation in contemporary societies across the world. The series combines the search for local, qualitatively rich accounts of everyday leisure with the international reach of debates in politics, leisure and social and cultural theory. In doing this, we will show that critical studies of leisure can and should continue to play a central role in understanding society. The scope will be global, striving to be truly international and truly diverse in the range of authors and topics.

Titles include:

Brett Lashua, Karl Spracklen and Stephen Wagg (*editors*)
SOUNDS AND THE CITY
Popular Music, Place and Globalization

Oliver Smith
CONTEMPORARY ADULTHOOD AND THE NIGHT-TIME ECONOMY

Karl Spracklen
WHITENESS AND LEISURE

Robert A. Stebbins
CAREERS IN SERIOUS LEISURE
From Dabbler to Devotee in Search of Fulfilment

Soile Veijola, Jennie Germann Molz, Olli Pyyhtinen, Emily Höckert and Alexander Grit
DISRUPTIVE TOURISM AND ITS UNTIDY GUESTS
Alternative Ontologies for Future Hospitalities

Leisure Studies in a Global Era
Series Standing Order ISBN 978–1–137–31032–3   hardback
978–1–137–31033–0   paperback
(*outside North America only*)

You can receive future titles in this series as they are published by placing a standing order. Please contact your bookseller or, in case of difficulty, write to us at the address below with your name and address, the title of the series and the ISBN quoted above.

Customer Services Department, Macmillan Distribution Ltd, Houndmills, Basingstoke, Hampshire RG21 6XS, England

---

Disruptive Tourism and its Untidy Guests

Alternative Ontologies for Future Hospitalities

Soile Veijola
University of Lapland, MTI, Finland

Jennie Germann Molz
College of the Holy Cross, USA

Olli Pyyhtinen
University of Tampere and University of Turku, Finland

Emily Höckert
University of Lapland, MTI, and University of Helsinki, Finland

and

Alexander Grit
Stenden University and University of Leicester

First published 2014 by
PALGRAVE MACMILLAN

Palgrave Macmillan in the UK is an imprint of Macmillan Publishers Limited, registered in England, company number 785998, of Houndmills, Basingstoke, Hampshire RG21 6XS.

Palgrave Macmillan in the US is a division of St Martin's Press LLC, 175 Fifth Avenue, New York, NY 10010.

Palgrave Macmillan is the global academic imprint of the above companies and has companies and representatives throughout the world.

Palgrave® and Macmillan® are registered trademarks in the United States, the United Kingdom, Europe and other countries.

ISBN 978–1–137–39949–6

This book is printed on paper suitable for recycling and made from fully managed and sustained forest sources. Logging, pulping and manufacturing processes are expected to conform to the environmental regulations of the country of origin.

A catalogue record for this book is available from the British Library.

A catalog record for this book is available from the Library of Congress.

Typeset by MPS Limited, Chennai, India.

Contents

Acknowledgements

Disruptive Tourism and its Untidy Guests began its life as a writing camp titled *Camping Together: A Tourist Experiment in Post-Biopolitical Living*. The camping trip was arranged at Keropirtti, near the Pyhä Fell in Finnish Lapland, from 18 February through 3 March 2013. We owe a debt of gratitude to the community of people who contributed in various ways to our experiment in real life and in writing. To begin with, we wish to thank our fellow campers, Tim Edensor and Gavin Urie, for their intellectual and artistic generosity during the camp. We are also grateful to those who planned the camp with us but in the end were unable to join us at Keropirtti: Julian Reid, Petra Falin, Suvi Alt and Paul Lynch. We thank our driver Heikki Aakkonen for all the hospitable transportations.

For their careful proofreading or insightful commenting on individual chapters of our book, we thank Suvi Alt, Joop Bos, Lydia Brauer, Maria Bergerlind Dierauer, Richard Foley, Szilvia Gyimóthy, Hans Christian Hansen, Karin Henningsson, Maaike de Jong, Martijn Menken and Kaisa-Liisa Puonti. Warm thanks are due to Johan Edelheim and Heli Ilola at Multidimensional Tourism Institute (MTI) for lending a hand with the index. Andrew James, Beth O'Leary and the production team at Palgrave Macmillan have been a delight to work with, and we thank them for their faith and patience with our disruptive tendencies.

We are grateful for the financial support from the University of Lapland for Soile Veijola's project on future tourist communities, under the working title *Acapella Village*, which enabled us to arrange the writing camp. We wish to thank the Fulbright Center for sponsoring Jennie Germann Molz' visit to the University of Lapland during the spring semester of 2013, thereby making it possible for her to participate in the camp and the writing of this book. For the cover image, we thank Markku Akseli Heikkilä. Finally, we thank each other for being such jolly campers.

About the Authors

Jennie Germann Molz is Associate Professor of Sociology at College of the Holy Cross in Worcester, Massachusetts, where she teaches courses on travel and tourism, global culture and society, and technology and mobility. Her research focuses on questions of togetherness and belonging in the context of tourism mobilities, with an ongoing focus on the ethical dimensions of hospitality. She co-edits the journal *Hospitality & Society* and has contributed articles on tourism, hospitality, food and cosmopolitanism to journals including *Mobilities, Annals of Tourism Research, Body & Society, Citizenship Studies, Space and Culture* and *Tourism Geographies*. She is the co-editor of *Mobilizing Hospitality: The Ethics of Social Relations in a Mobile World* (2007) and the author of *Travel Connections: Tourism, Technology and Togetherness in a Mobile World* (2012).

Alexander Grit is Research Lecturer at Stenden University, The Netherlands. He is also a senior research fellow affiliated to the Managing Science, Technology and Knowledge Research Group (MSTK) at the University of Leicester, School of Management. Grit teaches courses on leisure studies, tourism development, urban development, hospitality and retail concept development, and conducts research with students and partners into the art of facilitating serendipity in spaces of hospitality. His research interests focus on the health of interactional spaces, art and hospitality, serendipitous processes and Deleuzian Philosophy. He has written book chapters and articles on art, retail concepts, home exchanges and museums to edited books such as *The Critical Turn in Tourism Studies: Creating an Academy of Hope* (2010), and journals such as *Research in Hospitality Management*.

Emily Höckert has worked as a researcher at the University of Helsinki in the research project POLITOUR: Policies and Practices of Tourism Industry – A Comparative and Interdisciplinary Study on Central America (Academy of Finland 2011–2014). She is a doctoral student in Tourism Studies at Multidimensional Tourism Institute (MTI) at the University of Lapland in Rovaniemi, Finland, and her thesis deals with subjects of welcome in rural tourism development settings.

Höckert teaches courses on international mobilities, rural tourism and cultural studies of tourism at both universities. She has published an article on social and cultural significance of community-based tourism in *The Finnish Journal of Tourism Research* and co-authored a chapter on responsible rural tourism in *Matkailututkimuksen lukukirja* (A Guide to Tourism Studies, 2013).

Olli Pyyhtinen is Professor of Sociology at the University of Tampere and Senior Lecturer in Sociology at the University of Turku, Finland. His main research interests relate to social theory, philosophy, science and technology studies as well as the study of art, and address issues such as the notion and metaphysics of the social, togetherness, materiality, the gift and the constitution of the human. He is the author of *Simmel and 'the Social'* (2010) and *The Gift and Its Paradoxes* (2014). Pyyhtinen has contributed chapters to edited volumes, and his articles have appeared in journals including *Theory, Culture & Society, Continental Philosopher Review, Anthropological Theory* and *Process Studies and Distinktion*.

Soile Veijola is Professor of Cultural Studies of Tourism at MTI, University of Lapland, Finland. Her background is in sociology and she teaches courses on cultural and social studies of tourism, tourism as work, ethical epistemologies of tourism and academic writing. Her research deals with social production of knowledge, social relations, gender and embodiment; tourism work and future tourist communities. She has contributed chapters and articles, often with Eeva Jokinen, that examine tourism and hospitality from critical and feminist perspectives, in books such as *Touring Cultures, Visual Culture and Tourism, Travels in Paradox* and *Real Tourism,* and in journals such as *Theory, Culture & Society, NORA, Tourist Studies* and *Annals of Tourism Research*.

1
Introduction: Alternative Tourism Ontologies

We are expecting guests so we are cleaning our house – even though we very well know that after our guests leave, the house will likely be even messier than it was before. What is the point then? Is all the trouble for nothing? No. If the house was too dirty, our guests would not be able to enjoy themselves, no matter how good the talk, the food, the coffee or the wine. Or, even worse, if they knew our house was always a mess, they would stay away. No one would come. Indeed, should we ever want to *sell* our home we would take the deep cleanse even further: we would hide photographs, piles of paper, pieces of clothing – all belongings that are too personal – in the cupboards, in the attic or in the cellar. In that case, we would make our home into a blank screen onto which potential buyers could project future scenarios of themselves hosting their own guests in this space. For now, however, we tidy up just enough to make our guests feel welcome, but not so much as to erase every trace of ourselves.

Dirt is the mark of ownership – *'appropriation takes place through dirt'*.[1] The dirtier one's space is, the more the person is attached to it, and the less accessible it is to others. Hospitality, on the other hand, is closely tied with cleanliness. 'Purifying one's space is an act of welcoming', proposes philosopher Michel Serres.[2] In other words, hospitality is the opposite of appropriation: while appropriation is to take what is public and common and make it one's own by way of fouling or enclosing, hospitality means opening up one's private property and transforming it into something public and accessible for others.

While it is by leaving one's stain and disorder that one makes something one's own, one welcomes others by cleaning oneself and one's space. It is by removing, dispelling or hiding the marks of personal possession that we make our space available for others. In order to be able to *give space* to the guest the host literally needs to *clear up space* so that the other can arrive.

And what about the guest? How tidy must the guest be in order to stay or to be welcomed in the first place? Unless we are aiming to practise philosopher Jacques Derrida's 'absolute hospitality'[3] – an unconditional welcoming of the other, regardless of who that other may be – there are social rules, regulating and constitutive ones, pertaining to being a guest. For instance, by welcoming a guest into our home, we tend to impose on her or him the house rules or the horizons of expectations that this particular shared social space presupposes. From national borders to the thresholds of private homes, we may find our welcome – the one we extend, or the one we receive – hangs in the balance of formulaic calculations of risk, difference and tidiness. Do we measure up? Are we the 'right' kind of visitors – the ones with money to spend and with return airplane or train tickets that will take us back where we came from when our planned stopover ends? Are we the 'right' kind of hosts? The ones with clean sheets and clean streets? What is lost in these calculations? What do we miss out on when tidiness becomes a prerequisite for the welcome?

Welcoming untidy guests

This book argues that tourism need not merely cater to the paying client, but that it must also welcome *untidy guests*. Rather than thinking of the untidy guests in empirical categories, we are speaking of them metaphorically and symbolically. Untidy guests are the ones who give an unexpected twist to the social situation we live in, as well as the human condition in general, and who, ultimately, make life worth living and the world liveable. In a parallel fashion, tourism also needs to accept *untidy hosts* who live their lives instead of becoming the animators of it, turning invisible at the sight of a guest. This is not to suggest, however, that guests and hosts stay neatly on their sides of the dichotomy. Indeed, one of the untidy things that we must also come to terms with is the blurring between hosting and guesting as performances and experiences,[4] as hosts become guests and guests become hosts. Instead of crossing out the otherness of the other and forming community only among those with whom we appear to have something in common, we need to configure hospitality also in relation to the unknown, uninvited and possibly disturbing stranger who visits us, lives next door – or is within us.[5]

Why talk about untidiness? After all, is tourism not about preparing spaces, places and locations to be as shiny, comprehensible and visitable[6] as possible for anyone coming by? Is it not about making hotel beds, grooming ski slopes, shoring up beaches with sand, mowing golf courses, levelling slums and organizing displays of indigenous cultures before the guests arrive?[7]

For us, all this tidying up – no matter how well intentioned – sweeps the generative possibilities of tourism under the rug and ensures that we will continue to repeat existing patterns of governance and inequality. We hold the view in this book that when we are confronted with the unexpected, the unfamiliar or the illegible we can no longer affirm our old ways of thinking, feeling and acting, but have to find alternative, perhaps radical, ways of connecting with others, ourselves and our environments. A different affect, as philosophers Gilles Deleuze and Felix Guattari[8] would put it, comes into play at such moments, with the potential to suspend fear, violence and negligence, and replaces them with more ethical ways of relating to unknown others and ourselves in new situations. Instead of trying to bring order[9] to tourism by stage-managing or developing it – as the majority of tourism and hospitality course books would have it – we argue that theorists, planners, activists and tourists alike should embrace the messiness of hospitalities in tourism.

Thus *Disruptive Tourism and Its Untidy Guests* offers the messy metaphor and vital image of 'untidiness' for thinking about alternative arrangements of future hospitalities. For this purpose, we play with the counter-intuitive notions of *camping, parasite, silence, unlearning* and *serendipity* in our book. With the help of these 'untidy' concepts, we wish to disrupt the ontological presumptions of tourism and its theories, which tend to emphasize western notions of managerial order in, among other things, modernist development projects in the so-called Global South.

It is worth adding that we are not *against* the abiding trend in tourism studies towards more sustainable alternatives to the largely unequal capitalist structure of global tourism.[10] We share many of these aims, but we argue that tourism and hospitality studies should focus on *disrupting* – rather than sustaining – the taken-for-granted composition of tourism and hospitality, and their overriding concerns of management and capitalization of social and communal relations. Thus, we share the views presented within the critical studies of tourism where the need to disrupt dominating discourses of managerialism and neoliberalism in research has also been forcefully articulated.[11]

However, instead of continuing to valorize the pressing expectations of enhanced self-reflexivity within the Tourism Academia in general terms, we focus in our book on *potentialities of being-with other people*, with less (not more) organization and management, taking 'withness' or togetherness as the ontological starting point of life in general and tourism in particular. For philosopher Jean-Luc Nancy, for instance, being always appears as being-with; existence is necessarily and essentially co-existence.[12] On this basis, we configure, in what follows, *alternative ontologies of tourism* to the ones that take reality to exist through clear-cut and self-subsistent beings, subjects and categories. For imagining alternative ontologies, we introduce the concept of 'the untidy guest' to argue that when scholars – and, indeed, tourism itself – confound and interrupt habitual interactions and assumptions, this may lead to new ideas and understandings of the 'good life'.

Before discussing our paths to the unknown in more detail let us look into current understanding of links between disruptions, hospitalities and tourism as these have been transformed by the era of mobilization, and then return to the company of the untidy guest.

Disrupting tourism

When thinking about tourism and disruption, several images may come to mind. An initial thought might be the unexpected or undesirable disruptions that unsettle the current arrangements of tourism: high-season hurricanes; tsunamis; epidemic illnesses; volcanic eruptions; economic downturns; earthquakes; and terrorist attacks.[13] Another might be the ways in which commercialized mass tourism has been seen to disrupt local societies and economies.[14] In both cases, disruption motivates theorists, developers and planners alike to build more flexible products, services and workforces that will be resilient enough to weather the storms of economic or environmental upheaval. While acknowledging the way disruption intersects with tourism in the former cases, we are also concerned with disrupting the existing field of tourism and hospitality studies.

What, why and how exactly do we want to disrupt it? For starters, we think that the timeless question of *the host-guest relation* needs continuous rethinking. The 'guest situation', to use Helmuth Berking's[15] phrasing, has always been precarious. Taming the potentially life-threatening arrival of the stranger into a ritual that upholds and celebrates the local way of life has been one of the greatest social inventions in the history of the humankind. The ritualistic welcome also gave the host the power to define the guest situation.[16] The interesting etymological aspect of having the same root (*hospes, hostis* and *potis*) for enemy, guest and host tell a story beyond comparison in human history.[17] And yet, while the figure of the stranger is, in a way, the opposite of and threat to the moral and symbolic order of the local community, society relies on guests and strangers for its very being.[18] But the relation is only ever a temporary one, and only for a limited time. The duty of the stranger-as-guest is, as said earlier, to keep on moving after having held up a flattering mirror for the host.

There are other origin stories for the invention of hospitality out there as well,[19] but suffice it to say that hospitality's role has radically changed since the old days of treating any stranger equally, be it a beggar or a king. The same goes with the other human values: gift, justice and friendship. As Derrida has argued, hospitality, along the other values, will lose its meaning *as a value* as soon as it is offered in terms of reciprocal exchange, duty, debt or economy.[20] Nowadays, all of the aforementioned values are, as we know, mostly part of an exchange of some kind rather than 'pure gestures'; particularly modern tourism is a melting pot of values turning into commodities to be sold and purchased.

In a parallel manner, working life in wider terms has also turned into a market of exchanging hospitalities: engaging companies, associations, states and professionals in different branches to negotiate their positions by way of bonding, amenities and exchanges of hospitalities[21] – through affects and practices that used to mark the personal sphere. In sharp contrast, the millions of people without passports in camps in no man's land between countries[22] are a living proof that a human value cherished since primordial times, that of hospitality, is quite conditional and political[23] after all – in addition to being for sale in the market.

In this situation, one can imagine the difficulties in finding a fair and concise textbook definition of hospitality for the students of tourism and hospitality studies, or in initiating an ambitious scholarly debate on it. Given the contradictions related to the ethics and politics of hospitality, education and scholarship in relation to hospitality face a significant ontological challenge: how is the reality of *being* – and especially *being-with* – conceived when producing knowledge about hospitality in tourism? How *should* it be conceived? How *could* it be conceived? An easy way out of it is, of course, to narrow the focus to 'a cluster of service sector activities associated with the provision of food, drink and accommodation'.[24] We have, however, something else in mind. We want to disrupt the very starting points of thinking hospitalities in and outside tourism.

In our view, disruption is not something that happens to an already existing social order; it is what makes social order – and alternative social orders – possible.[25] In reality, disruption and sustainability are thoroughly relational, co-constituting elements of social life. This is why sociologist Georg Simmel conceives of the stranger who arrives as a *necessary* disruption. We, as individuals and as society, have to relate somehow to the stranger/guest in order to exist and form a group.[26] The social is constituted in the play between inclusion and exclusion. In other words, 'social order is consolidated by reintegrating that which disrupts it.'[27] Following Simmel, we do not see disruption as a failure of the social order, but rather as something that must be embraced in order to imagine and enact alternative possibilities for social life. In revealing the contingencies and vulnerabilities of our everyday social order, disruption can become an opportunity for arranging life differently.

In tourism studies, we wish to trouble the field's complicity with the status quo by emphasizing the generative and creative potential of disruption. This said, we do not suggest that either disruption or sustainability are qualities that can be objectively measured or evaluated. What appears to be sustainable from one perspective may very well look disruptive from another.

But given the existing (not merely theoretical) messiness of tourism, many of the calls for sustainable tourism focus on managing, developing, governing or policy-writing 'untidy guests' – as well as 'untidy hosts' – out of the picture. Simultaneously tourism students are often encouraged to memorize ready-made definitions of, say, hospitality management, heritage tourism or sustainable tourism development, instead of critically reflecting on the ontological premises of previous topics. In other words, the focus of tourism research has been laid extensively on the conditionality, limits and laws of hospitality while trying to close the door on the incalculable. The unexamined ontologies of many lines of tourism studies leave little room for thinking and doing togetherness between and among hosts and guests differently.

Alternative ontologies of tourism

In our book we treat *ontology* not as an already-set arrangement of the basic furniture of the world, but in an anti-foundational way that asks how reality is constructed and articulated in practice. Instead of starting from general categories and basic principles, which tell us in advance what the world is made of, we begin in the middle of things and lay out the world in all its messiness, openness and fickleness. And because we do not believe that there is only *an* ontology which would apply to each and every specific case, we subscribe to ontological pluralism.[28]

Moreover, we do not understand concepts as merely 'theoretical' or linguistic. As Deleuze and Guattari suggest, albeit being 'incorporeal', concepts are 'incarnated or effectuated in bodies'.[29]

In addition to their realness in our embodied lives, concepts can also be invented, following Deleuze's understanding of philosophy.[30] For him – and us – concepts are not merely representative or descriptive, but active and creative. In Deleuze's thought, philosophers are creators of concepts, while artists create affects and percepts, and scientists produce functions; accordingly, each philosophical encounter ought to inspire new concepts.[31] In our book, we, too, work across philosophy, science and art – and carve routes in an unknown territory or cut through our chaos in order to create, rather than manufacture,[32] knowledge.

Creating concepts therefore means potentially creating new possibilities for being and acting. It means renewing our practices; 'a new idea introduces a new alternative', as Alfred North Whitehead notes.[33] That is, we can affect and renew the existing arrangements of social relations – with disruptive tourism. Notwithstanding the undoubtedly critical ethos that the metaphors of untidiness and disruption announce, our primary aim in this book is *not* deconstructive or destructive but constructive. (Being constructive is fitting also because its prefix 'con-' implies 'withness'.) Philosopher Graham Harman has suggested that there are basically two kinds of criticism: that which wishes the interlocutor to fail and that which wishes her or him to succeed.[34] The first could be labelled by the word 'polemic'. The person the polemicist critiques, as philosopher and social historian Michel Foucault has noted, is not treated as 'a partner in the search for the truth but an adversary, an enemy who is wrong.'[35] The polemicist thinks that he (or, in some rare cases, she) alone has an immediate access to truth, and therefore he looks for where others have made mistakes, collects the proof of their guilt, pronounces the verdict and sentences them.[36] Polemic thus mimics war and battle: it is about annihilation, about destroying one's enemy, about crushing one's adversary.[37] Polemic also never invents anything new.[38] On the contrary, it has 'sterilizing' effects that tend to hold interlocutors fast in their current views and positions.[39]

It is in terms of the latter, say, *vital* sense of being critical, that we use our metaphor of disrupting. We see our task to be to stimulate, not paralyze or end, discussion. Disrupting is for us about bringing something – a concept, an idea, or a practice – alive,[40] about setting free what lives.

It bears to add that in the medical lexicon, the term *critical* describes a situation where the existence of an organism is in jeopardy, because of a stroke, a heart attack, an accident or an infection, for example. In such a situation, in order to survive and recover, the body is forced to reinvent its organization anew.[41] Then again, for Deleuze, the process of disruption means that in a process of *divergent actualization* the process of becoming is rather unpredictable; in a process of *repetition*, for its part, what one turns out to be or what happens is rather predictable.[42] Processes of repetition also repeat socio-economical inequalities and stratifications. In order to stop the process of reproducing sameness and excluding difference, a disruption is needed. Analogous to the previous claims, what we intend by the idea of being critical in a vital manner refers to the need to reinvent our lives and future. We aim to disrupt the current arrangements of social life and of hosting and guesting relations by welcoming disruption because of its ontological potentialities.

Camping together

The chapters that follow have emerged out of our experiments with camping. Our theoretical experiment has been inspired – and troubled – by philosopher Giorgio Agamben's notion of the camp as a state of permanent exception and thereby legitimate violence.[43] Aiming to engage with but also rework some of the negative connotations of Agamben's camp,[44] we have set out to 'camp' within and across territories of academic theorizing, discussing the possibilities of disrupting bio-politics. In this sense, our camping together has been more like the camping trip G. A. Cohen describes in *Why Not Socialism?* – an experiment in sharing relations and resources in the spirit of communality.[45] Knowing that tourism has been dictated by the notion of sustainability, but even more by neoliberalism which 'extends market rationality across the entire social field into the deepest crevices of individuality',[46] we have tried to *imagine tourism post-bio-politically* – whatever that could mean – and find ways of opening our minds for the unexpected. We have wanted to stir political and ethical imagination in envisioning alternative ways of *being*.[47] Our aim as an author collective has thus been to 'camp together' in order to create a more imaginative approach to sharing research ideas and creating a writing community in tourism and hospitality studies than is perhaps customary in the Academia of today.

We have also camped together in real life: for a week in early spring 2013 at a log-house in Finnish Lapland.[48] Our book thus has its roots in the conversations that unfolded during an unorthodox, academic camping trip – rare as it is in the academic life to spend more than two conference days together. Along with the idea of writing a book together – engendered in the morning of the last day at the camp and as a result of ongoing conversations about Jørgen Leth's and Lars von Trier's film *The Five Obstructions* (2003) – came the desire to challenge the standard templates of creating and writing scholarship in tourism and hospitality studies.[49]

In our book we play in both content and form with the idea of obstructions, which we have set for ourselves in the form of 'writing rules'. We start from the premise – deeply informed by the ethos of *The Five Obstructions*, which demonstrates how creativity 'feeds on limits'[50] – that rules can produce new possibilities. Namely, von Trier challenged his friend and mentor in film-making, Jørgen Leth, to remake Leth's 1967 film *The Perfect Human* five times, each time adhering to a different 'obstruction' devised by von Trier to confound his friend. Instead of constraining Leth's cinematic genius, the obstructions, or rules, result in ever more brilliant renditions of his original film.[51]

Inspired by their collaboration, our book thus follows a new set of five rules in order to break the rules of conventional knowledge production in tourism studies.[52] We require each of our chapters to: (1) invite two or more 'theoretical houseguests' (well-known philosophers, scholars and theorists) who do not address the world from a similar embodied speaking position; (2) disrupt established disciplinary silos of academic writing so that the book travels across existing academic boundaries; (3) care for the reader's affective experience through our narrative style; (4) position tourism studies not as an innocent bystander reporting on a changing world, but as an agent of change in the world; and (5) reflect the way we 'messed around' with each other's ways of thinking during our shared camping and writing experience. The task of the reader is however not to evaluate the extent to which we followed the rules but – perhaps – to play along in the alternating settings, finding the part to her or his liking to play.

We have not been alone in our quest. Other theorists and theories accompany us in this endeavour. In addition to the footprints of the fellow campers on our chapters, our book hosts conversation among an assemblage of well-known thinkers: Giorgio Agamben, Gaston Bachelard, Gilles Deleuze, Jacques Derrida, Paulo Freire, Elizabeth Grosz, Felix Guattari, Martin Heidegger, Luce Irigaray, Emmanuel Levinas, Jean-Luc Nancy, Friedrich Nietzsche, Michel Serres, Georg Simmel and Gayatri Chakravorty Spivak. The forays into philosophical and sociological thought, instead of engaging mainly with familiar discourses within tourism and hospitality studies, are meant to open up space for thinking differently, more creatively and ethically, about future tourism rather than merely contributing to naïve or predatory notions about the growth of the tourism industry.

This chapter, the introduction, is co-written by all of us. The next five chapters, Chapters 2 through 6, are individually designed and composed – but regularly looked into by other campers during the book-making process. The sequential order of the chapters is intuitive rather than cumulative; hence the book can be read in any order without a risk of falling off the cliff. The chapters address real-life sites, which can also be fictional: a camp site, a secluded tropical island, a familiar neighbourhood, coffee trails or an open air museum. (We do not write tourism geographies or area studies.) As literary campers entering these sites, we carry theoretical concepts that are not commonly seen together: *camping* and *unfinishedness, paradise* and *parasite, silence* and *community, hospitality* and *unlearning,* and *serendipities* and *hospity.* We do not yet know all the consequences of bringing them together!

In Chapter 2, Jennie Germann Molz explores camping as a tourist practice and as an alternative ontology for tourism. The camping itinerary ranges from memories of her own family camping trips to television representations of camping to *Cook It Raw,* an annual camp for the world's greatest chefs. Germann Molz interprets these camping experiences in light of Heidegger's concept of clearing, Grosz's discussion of *chora* and Freire's notion of unfinishedness. She concludes that clearing, *chora* and unfinishedness all gesture towards the emptiness and incompleteness that allows for serendipity in tourist encounters, opening up new possibilities for the future of tourism. In her view, we often mistake fullness as the precondition for sociability in tourism, when in fact it is emptiness and the state of unfinishedness that makes moves – towards and away from one another and into altogether new configurations – possible in the first place. The chapter contributes to the book's aim to develop ontology for future tourism by offering ways of thinking about new places and arrangements of being and being-with in tourism encounters.

Olli Pyyhtinen in Chapter 3 takes up from Serres the untidy notion of 'the parasite'. The concept has quite a dispersed set of referents: an organism feeding on another; an abusive or uninvited guest who takes without giving anything in return; or noise, static, a break in the message. Pyyhtinen uses it to examine hospitality and the constitution of community through the interplay of inclusion and exclusion. The chapter makes philosophy and theory bear upon fiction, as it discusses these themes by engaging with the film *The Beach* directed by Danny Boyle in 2000. Set in Thailand, the film is a travel story of the pursuit of a perfect beach, a paradise on earth. Pyyhtinen suggests that the movie displays how every paradise, whether in the form of pristine nature or an ideal community, is itself a product of violence, as it is only made possible by way of excluding parasites. At the same time, however, the total elimination of parasites cannot be attained – parasites keep turning up, no matter what. Ultimately, the chapter proposes the idea that in order to prevent hospitality from turning into hostility, the community must make its peace with the parasites, and that thereby the truce with the parasite may present a very condition of ethics itself.

Soile Veijola in Chapter 4 builds an understanding of silence as the absence of interruptions and a source for an ethical ontology of being-with-many, drawing inspiration from especially Nancy and Simmel. With the help of a real-life event of interactive environmental art, *The Reindeer Safari*, and the notion of silence she disrupts and revitalizes the ideas of 'the social', community, and hospitality, and, finally, elaborates on the notion of the silent communities as a form of mobile neighbouring. Involving also D. W. Winnicot, Mihaly Csikszentmihalyi and Iris Marion Young in her theoretical itinerary, Veijola questions the ways in which 'one' and 'many' as well as 'being-two' are constituted in our discourses of tourism and hospitality and contemplates on the possibilities of addressing plurality as embodied being-withness in ethically and ontologically consequential ways. The silent communities are offered as an example of ethical plural for the future tourism hospitalities.

In Chapter 5, Emily Höckert examines the ethics of togetherness by paying attention to the relations of welcoming between, on the one hand, rural communities as hosts and, on the other hand, tourism researchers and developers as 'tidy guests'. The chapter presents a fictional field visit to Northern Highlands of Nicaragua weaving together pieces of empirical data with Levinas', Derrida's and Spivak's theories of ethical subjectivities. Drawing on Spivak, Höckert argues that the perceived privileged position of the mobile guests justifies interpreting, speaking for and speaking about the other, which, for its part, leads to epistemic violence: to erasure and trivialization of other ways of knowing the world. In the chapter, the notion of hospitality, 'welcome of the other', lends itself as an analytical and reflective tool for 'learning to unlearn' one's privilege as loss and for envisioning ethics as a desire of infinity and receptivity. By her emphasis on 'being-for-the-other', she underlines the possible limitations and ethical, political implications of being a tidy guest and encourages one to reflect one's openness towards the other.

Alexander Grit, in Chapter 6, examines the creative potentialities and limits of tourism and hospitality through the notion of the untidy guest, which for him is made of serendipities and becomings. In Grit's thought, both host and guest can alter their experience of hosting and guesting towards a state of *hospity*, a concept he elaborates on in an embodied fashion by visiting the open air museum Zuiderzee museum with his daughter and his friend Kwame. Grit criticizes fixed spaces of hospitality and argues for spaces of hospitality whereby both host and guest can construct their own trajectories of exploration. His approach is a vitalist one, allowing for addressing spaces of hospitality from the point of view of their healthiness, thereby turning the social scientist into a physician of culture of some sort. The chapter provides criteria for assessing health in spaces of hospitality on the basis of whether host and/or guest engage in unexpected findings and whether these findings can be integrated within rather predictable interactions. By messing around with the concept of *hospity*, Grit creates spaces of hospitality where the roles of the host and the guest can alternate and play with future potentialities.

The final chapter is a co-written conclusion in which we take one more round in the theoretical landscape of the book and comment on the *prepositions* that perform invisible yet important work in our propositions for a set of alternative ontologies for future tourism hospitalities.

***

The guests might stay longer than we had expected. We are starting to get annoyed by all the things that are not where they used to be. The children leave their toys in the wrong places. Adults are reckless with their clothes, towels and beer cans. They have taken too many books off the bookshelf and left them who-knows-where. Two *Lego*-people are in the doll-house, squeezed onto a miniature sofa. A plastic dinosaur stands in front of them, blocking their view of the tiny wooden television. We feel the urge to start returning the toys to their proper places; to begin collecting the half-full glasses from our guests' hands so that we can run the dishwasher. Or maybe it would be easier and faster to organize everything back to normal after the guests have left? But could the *Lego*-people and the dinosaur stay put in the doll-house? Could the pillows find new places? Could they even be sat on? Why do we have to hurry to restore the old order? Might we be missing, even losing, something while worrying about organizing everything back to what its former order? Could the mess be a start of something new? For no matter how hard we worked to bring order back, things can never be restored to their previous stage. There is no 'return' to the earlier condition. No one and nothing leaves the hospitality assemblage unchanged; the host, the guest, their relation as well as the natural environment and the material artefacts are transformed, sometimes radically, sometimes only to an almost imperceptible degree. They are given new properties. Untidiness thus produces and indicates a new state, a space of novelty and creation. Overall, what if untidiness would not be perceived as 'a state of exception', as Agamben would put it, but as the 'new normal', the unfinishedness of being-with?

An untidy guest turns hosting into an ethical choice. An untidy host(ess) does the same to the guest. Both give the saying 'you are a mess' a positive tone. Who knows, the *dirt* could, after all, be in its proper place, and thus cease to be dirt – as Mary Douglas[53] might end this conversation. Quite uninvited, we must say.

Notes

1. Michel Serres, *Malfeasance. Appropriation through Pollution?*, translated by Anne-Marie Feenberg-Dibon, Stanford, Stanford University Press, 2011, p. 3; italics in the original.
2. Michel Serres, *The Parasite*, translated by Lawrence R. Schehr, Minneapolis, University of Minnesota Press, 2007, p. 145.
3. Jacques Derrida, *Adieu to Emmanuel Levinas*, translated by Pascale-Ann Brault and Michael Naas, Stanford, Stanford University Press, 1999.
4. David Bell, 'Moments of Hospitality', in Jennie Germann Molz and Sarah Gibson (eds), *Mobilizing Hospitality: The Ethics of Social Relations in a Mobile World*, Aldershot, Ashgate, 2007, pp. 29–45.
5. See Sara Ahmed, *Strange Encounters: Embodied Others in Postcoloniality*, London, Routledge, 2000; Molz and Gibson, *Mobilizing Hospitality*; Gideon Baker, *Politicising Ethics in International Relations: Cosmopolitanism as Hospitality*, London, Routledge, 2011; Rauna Kuokkanen, *Reshaping the University: Responsibilities, Indigenous Epistemes, and the Logic of the Gift*, Vancouver, University of BC Press, 2007.

6. See Bella Dicks, *Culture on Display: The Production of Contemporary Visitability*, Maidenhead, McGraw Hill, 2004.

7. For example, see Nigel Morgan and Stroma Cole (eds), *Tourism and Inequality: Problems and Prospects*, Oxfordshire, CABI, 2010; Solomon Greene, 'Staged Cities: Mega-events, Slum Clearance, and Global Capital', *Yale Human Rights and Development Law Journal*, 6, 2003, pp.161–179; Rosaleen Duffy, *A Trip Too Far: Ecotourism, Politics, and Exploitation*, London, Earthscan, 2002; Simon Romero, 'Slum Dwellers Are Defying Brazil's Grand Design for Olympics', *The New York Times*, 4 March 2012.

8. See, for example, Gilles Deleuze and Felix Guattari, 'Third Aspect: The War Machine and Nomad Affects', in *A Thousand Plateaus: Capitalism and Schizophrenia*, translated by Brian Massumi, Minneapolis, University of Minnesota Press, 1987, pp. 327–328.

9. Adrian Franklin, 'The Tourism Ordering. Taking Tourism More Seriously as a Globalising Ordering', *Civilisations*, LVII.1–2 – *Tourisme, mobilités et altérités contemporaines*, 2008.

10. For an in-depth view on the subject of sustainability in tourism, see Jarkko Saarinen, 'Traditions of Sustainability in Tourism Studies', *Annals of Tourism Research*, 33.4, 2006, pp. 1121–1140; David Weaver, Ralf Buckley, Brian Wheeller and Bill Bramwell, 'Mass Tourism and Sustainability: Can the Two Meet?', in Tej Vir Singh (ed.), *Critical Debates in Tourism*, Bristol, Channel View Publications, 2012, pp. 27–52.

11. For example, Richard Sharpley, *Past Trends and Future Directions*, Oxon, Routledge, 2011; Annette Pritchard, Nigel Morgan and Irena Ateljevich, 'Hopeful Tourism, a New Transformative Perspective', *Annals of Tourism Research*, 38.3, 2011, pp. 941–963; Freya Higgins-Desbiolles, 'More Than an "Industry": The Forgotten Power of Tourism as a Social Force', *Tourism Management*, 27.6, 2006, pp. 1192–1208.

12. Jean-Luc Nancy, *Being Singular Plural*, translated by Robert D. Richardson and Anne O'Byrne, Stanford, Stanford University Press, 2000.

13. Thomas Birtchnell and Monika Büscher, 'Stranded: An Eruption of Disruption', *Mobilities*, 6.1, 2011, pp. 1–9.

14. For example, see Morgan and Cole, *Tourism and Inequality*; Valene Smith (ed.), *Hosts and Guests: The Anthropology of Tourism*, Philadelphia, University of Pennsylvania Press, 1989; Erik Cohen, 'The Impact of Tourism on the Hill Tribes of Northern Thailand', *Internationales Asienforum*, 10.1/2, 1979, pp. 5–38.

15. Helmuth Berking, *Sociology of Giving*, translated by Patrick Camiller, London, Sage, 1999, p. 82.

16. Ibid., p. 92.

17. Émile Benveniste, *Indo-European Language and Society*, translated by Elizabeth Palmer, London, Faber and Faber Limited, 1973.

18. Chris Rumford, *The Globalization of Strangeness*, Chippenham & Eastbourne, Palgrave Macmillan, 2013.

19. See, for example, Paul Lynch, Jennie Germann Molz, Alison McIntosh, Peter Lugosi and Conrad Lashley, 'Theorising Hospitality', *Hospitality &*

Society, 1.1, 2011, pp. 3–24; Conrad Lashley and Alison Morrison (eds), *In Search of Hospitality: Theoretical Perspectives and Debates,* Oxford, Butterworth-Heinemann, 2000; Kevin O'Gorman, 'Discovering Commercial Hospitality in Ancient Rome', *Hospitality Review,* 9.2, 2007, pp. 44–52; Judith Still, 'France and the Paradigm of Hospitality', *Third Text,* 20.6, 2006, pp. 703–710.

20. Jacques Derrida, *Of Hospitality: Anne Dufourmantelle Invites Jacques Derrida to Respond,* translated by R. Bowlby, Stanford, Stanford University Press, 2000; See also Sarah Gibson, 'Accommodating Strangers: British Hospitality and the Asylum Hotel Debate', *Journal for Cultural Research,* 7.4, 2003, pp. 367–386.

21. Soile Veijola, 'Gender as Work in the Tourism Industry', *Tourist Studies,* 9.2, 2009, pp. 109–126.

22. Paul Virilio, *The Futurism of the Instant. Stop-Eject,* Cambridge, Polity Press, 2010.

23. Derrida, *Of Hospitality,* pp. 21–25.

24. Conrad Lashley, 'Towards a Theoretical Understanding', in Lashley and Morrison (eds), *In Search of Hospitality,* p. 2.

25. On this point, see Jonathan Paul Marshall and James Goodman, 'Disordering Network Theory: An Introduction', *Global Networks,* 13.3, 2013, pp. 279–289, who argue against a 'prevailing and historical bias towards order' and insist, instead, that we must embrace a 'paradox of disorder/order' that sees disruption as integral to the very systems it disrupts.

26. Georg Simmel, Soziologie, in Georg Simmel, *Gesamtausgabe, Band 11,* Frankfurt am Main, Suhrkamp, 1908/1992.

27. Sergio Benvenuto, 'Fashion: Georg Simmel', *Journal of Artificial Societies and Social Simulation,* 3.2, 2000.

28. For more on ontological pluralism, see Bruno Latour, *An Inquiry into Modes of Existence: An Anthropology of the Moderns,* translated by Catherine Porter, Cambridge, MA, Harvard University Press, 2013.

29. Gilles Deleuze and Felix Guattari, *What is Philosophy?,* translated by Hugh Tomlinson and Graham Burchell, New York, Columbia University Press, 1994, p. 21.

30. Ibid.

31. Ibid., p. 163.

32. Donna Haraway, 'A Cyborg Manifesto: Science, Technology, and Socialist-Feminism in the 20th Century', in *Simians, Cyborgs and Women: The Reinvention of Nature,* New York, Routledge, 1991, p. 183.

33. Alfred North Whitehead, *Process and Reality,* Corrected Edition, New York, Free Press, 1929/1978, p. 11.

34. Graham Harman, *The Prince of Networks: Bruno Latour and Metaphysics,* Melbourne, Re-Press, 2009, p. 119.

35. Michel Foucault, 'Polemics, Politics and Problematizations', in Paul Rabinow (ed.), *Essential Works of Foucault 1954–1984: Ethics, Subjectivity and Truth,* London, Penguin Books, 2000, p. 112.

36. Ibid.

37. Ibid., p. 113.

38. See Serres in Michel Serres with Bruno Latour, *Conversations on Science, Culture, and Time*, translated by Roxanne Lapidus, Ann Arbor, University of Michigan Press, 1995, p. 38.
39. Foucault, 'Polemics, Politics and Problematizations', p. 113.
40. See Michel Foucault, 'The Masked Philosopher', in Paul Rabinow (ed.), *Essential Works of Foucault 1954–1984: Ethics, Subjectivity and Truth*, New York, New Press, 2000, p. 323.
41. Michel Serres, *Times of Crisis: What the Financial Crisis Revealed and How to Reinvent Our Lives and Future*, translated by Anne-Marie Feenberg-Dibon, New York and London, Bloomsbury, 2014, pp. xi–xii.
42. Gilles Deleuze, *Difference and Repetition*, translated by Paul Patton, New York, Columbia University Press, 1994.
43. Giorgio Agamben, *Homo Sacer: Sovereign Power and Bare Life*, translated by Daniel Heller-Roazen, Stanford, Stanford University Press, 1998.
44. See, for example, Ibid.; Bülent Diken and Carsten Bagge Laustsen, *The Culture of Exception: Sociology Facing the Camp*, London and New York, Routledge, 2005; G.A. Cohen, *Why Not Socialism?*, Princeton and Oxford, Princeton University Press, 2009; Claudio Minca, 'No Country for Old Men', in Claudio Minca and Tim Oakes (eds), *Real Tourism: Practice, Care and Politics in Contemporary Travel Culture*, London, Routledge, 2011, pp. 2–37.
45. Cohen, *Why Not Socialism?*
46. Wanda Vrasti, *Volunteer Tourism in the Global South: Giving Back in Neoliberal Times*, London and New York, Routledge, 2013, p. 32; see also Jason Read, 'A Genealogy of Homo Economicus: Neoliberalism and the Production of Subjectivity', *Foucault Studies*, 2006.6, 2009, pp. 25–36; David Harvey, *A Brief History of Neoliberalism*, New York: Oxford, 2005.
47. See Vrasti, *Volunteer Tourism in the Global South*, p. 31.
48. 'A writing camp' by the name *Camping Together: A Tourist Experiment in Post-Biopolitical Living*, in which the authors of this book participated, was arranged at Keropirtti, a retreat for researchers offered by the University of Lapland by the Pyhä fell, 18 February–3 March 2013. Olli Pyyhtinen was present through his book manuscript *The Gift and Its Paradoxes: Beyond Mauss* (Aldershot: Ashgate, 2014) which was being read by another camper during the camp. In retrospect, we like to think that ours resembles the writing process of the book *Writing Culture*, edited by George Marcus and James Clifford in 1986, which was also a camping experiment of its kind and at its time changed the scene of knowledge production in anthropology. We wish to make a similarly minded intervention in tourism and hospitality studies, without however expecting to become a similar milepost as *Writing Culture* was.
49. See, for example, Soile Veijola and Eeva Jokinen, 'The Body in Tourism', *Theory, Culture & Society*, 11.3, 1994, pp. 125–151.
50. For the suggestion to watch the film *The Five Obstructions* in the camp we are grateful to Julian Reid (personal communication).
51. The same idea has inspired other experiments with writing as well, such as those by the Oulipo group, founded in the 1960s. The Oulipo group

was a gathering of francophone writers who imposed on themselves various techniques of constrained writing in order to spark creativity and inspiration.

52. For examples of earlier experiments on breeching the narrative structures of academic writing, see Veijola and Jokinen, 'The Body in Tourism'; and Soile Veijola and Eeva Jokinen, 'The Death of the Tourist', *European Journal of Cultural Studies*, 1.3, 1998, pp. 327–351.

53. Mary Douglas, *Purity and Danger*, London and New York, Routledge, 1966/2003.

2
Camping in Clearing

Jennie Germann Molz

Introduction

Camping confounds the imaginaries of indulgence and comfort that glisten from the pages of travel and lifestyle magazines. The glossy layouts, filled with pictures of all-inclusive beach resorts and cruise ships, promise to fill the hollowness of modern lifestyles with an overabundance of just about everything. Instead of pampered luxury, however, *camping* embraces discomfort and lack – not so much as pleasures in and of themselves but as reminders that less is often more. This chapter is about those empty intervals in space and time that trigger our creative impulses by making room for us to move in different directions, to create unexpected connections, or to arrange our social worlds in new ways. I look to camping, as a practice and as a metaphor for life in an uncertain world, to explore concepts we rarely associate with tourism – concepts such as emptiness or unfinishedness. To do this, I first draw on the ontologies of Martin Heidegger's 'clearing' and Plato's *chora*, reading them through the critical lenses offered by feminist theorists Elizabeth Grosz and Betsy Wearing, before turning to Paulo Freire's notions of 'unfinishedness' to reflect further on alternative ontologies for tourism.

Arriving at the campsite and making ourselves at home in the emptiness that we find (or invent) there engenders a kind of creative making do, a making out of nothing, which resists – even as it taps into – the materialism of modern life. Camping reminds us that our material and social worlds are not inevitable, that they are contingent and provisional and that their tangible durability is always in the making. This matters, because new possibilities live precisely in the *evitability* of what we take for granted. In other words, camping is an ontological experimentation. Its embrace of emptiness and unfinishedness is an exercise in being and becoming. So this chapter is a camp story.

As far as stories go, however, it is by design an incomplete one. It is not a systematic analysis of the proliferating practices of modern camping, or of its theoretical consequences, but rather an assemblage of personal memories, bits of popular culture and philosophy. My approach follows Steven Shaviro and his interpretation of Whitehead's philosophy:

> Philosophical speculation collects the most heterogeneous materials, and puts them together in the most unexpected configurations. It is something like the practice of collage in modernist painting; or better – to use an analogy not from Whitehead's time, but from our own – it is like a DJ's practice of sampling and remixing.[1]

What follows is thus an attempt to extract 'patterned contrasts' from an assemblage of heterogeneous materials in order to tell a somewhat larger story about ontologies of camping and camping as ontology.[2] Nevertheless, the story assembled here and its patterned contrasts leave gaps, some of which are intentional.

There are dozens of iterations of the camp. In *Camps: A Guide to 21st-Century Space*, Charlie Hailey takes us on a tour of some of the more prominent ones:

> We camp with kids in our backyards, we arrange ourselves in partisan camps, we watch as camps overflow with twenty million refugees, we fill arenas with disaster victims, we speculate about the locations of terrorist camps, and we marvel at North America's burgeoning RV [recreational vehicle] culture.[3]

The list continues: boot camps, migrant camps, protest camps, base camps, summer camps, trailer camps, nudist camps, concentration camps, peace camps, Gypsy camps, climate camps and so on.[4] It's almost more than one word can take.

Hailey goes on to explain that the social and spatial logic of the camp – its delicate balance between freedom and control, temporariness and permanence, unfinishedness and durability – has become the model of what it is to live in 'itinerant times' and under conditions of 'collective homelessness'.[5] While his list of camps highlights several practices of recreational camping such as caravanning and summer camps, it is the image of a coil of barbed wire on his book's title page that strikes me most. The image is far more suggestive of spaces of control than of leisure. This is not surprising, given that much of the recent scholarship on camps stems from Giorgio Agamben's argument in *Homo Sacer* that the camp has become the 'nomos' of modernity.[6]

According to Agamben, the dehumanizing logic of the concentration camp is hopelessly entangled with that of the modern state. The state of exception that justifies the camp becomes the normalized condition of everyday life. From this departure point, the connotation of the camp remains deeply linked with concentration camps, internment camps, refugee camps and other juridico-political structures through which the state relegates some people to 'bare life', life stripped of its political and cultural forms. I want to focus instead on the voluntary and leisurely practices of camping and on camping as a metaphor for disruptive tourism of the future, which is the subject of our book. Although we will see later that some theorists have transported Agamben's paradigm of the camp into the world of tourism, the story I tell here does not follow that route, at least initially. This is part of the reason why it is incomplete. But only part of the reason.

In the pages that follow, I assemble three somewhat disparate examples of camping ranging from my own childhood memories of camping to television sitcoms and culinary experiments in the wild. My examples are loosely held together by meditations of a more philosophical nature, drawn around the ontological concepts of 'clearing' and *chora*, and unfinishedness, which I introduce as 'untidy guests' that must be welcomed rather than turned away.

Camp grounds

My childhood memories of family vacations are memories of camping. When summer rolled around, we would pile all of our gear – tents, sleeping bags, kerosene lanterns, bags of charcoal, bug spray – into our green Volkswagen van and head to some campground in the woods or out in the desert. Sometimes my grandparents or some assortment of aunts, uncles, cousins and friends would meet us there. I remember the ritual of pulling into our campsite, unloading everything and clearing away the pine cones and large rocks from the pads where our tents would go. I remember my dad wrestling with the metal skeleton of our canvas tents, and later with the elasticated supports of our nylon ones, and then securing our billowing tents to the ground with a mallet.

If camping is a metaphor for our collective homelessness, it is also a metaphor for our collective 'homing desire'.[7] As a child my first impulse on camping trips was not to revel in the suspension of social structures but rather to recreate the tangible and habitual anchors of hominess. The minute the tents were pitched, my sister and I would tumble into them to carve out the imaginary boundaries that transformed our tent into a home. 'This will be my bedroom and that can be the living room there', we would say as we traced with our fingers the unseen walls and doors that brought our interior space under control. We would unroll our sleeping bags and prop up pillows and position coolers and camp stools into never-quite-solid-enough markers of these domestic divisions. Most precarious was the invisible division between *my* bedroom and *her* bedroom.

Outside the tent, we would engage in a similar performance of homemaking, designating a wooden picnic table as the 'dining room', the charcoal grill as the 'kitchen' and the sandy pitch in front of our tents as the 'porch'. A broken branch protruding from a tree trunk could be a hook for a lantern or a hammock; we could balance some sticks like this to make a shelf; a clothes line pulled between two trees held a wilting wall of towels; a log became an armchair in front of the fire pit. Our playful efforts to make ourselves at home re-enacted the complex violences of creativity, exclusion and entitlement.

Our tent-pitching and homemaking were the opening rituals of camping, after which we would fall into familiar new routines of gathering firewood, playing cards on the picnic table, collecting blackberries, hunting fireflies. And of course roasting marshmallows over a smoky fire and pressing them into chocolate and graham cracker sandwiches. Holding long thin twigs into the fire and then waving their embered ends in the dark night air to spell our names in cursive. The adults on folding mesh lawn chairs drinking beer and telling stories, until too late into the night. In the mornings, the cloying smells of bacon and propane gas mingled with the fresh breeze. My mother would fill a plastic bucket with soap and water. The dishwasher.

Those are the rosy, mosquito-free memories. There are others. My father catching his shoe on fire trying to light the propane camp stove, huddling through a thunderstorm in a damp tent, my sister being rushed out of the woods to the hospital in the middle of the night for an emergency appendectomy or shivering through a too-cold night at Crater Lake in Oregon. Oh, and the RVs, the caravans, the trailers that would pitch camp nearby. The flickering blue glow from their windows evoking in me a mix of envy and disdain. We were purists. We built our home-away-from-home with tents and tarps and twigs; we didn't transport it wholesale in a plug-and-play box on wheels into the wilderness to watch TV.

I camp far less often now. Or perhaps I should say that I am camping all of the time now.

Clearing for camping

We camp in clearings. We clear a space for camping. We clear time for camping, too. Clearing refers to both the physical *topos* where we camp and to the temporality of camping. Clearing also refers to the act of making a place camp-able. It is a place, a time and a practice.

Is this what Martin Heidegger means when he writes about clearing? As Heidegger describes it, clearing is the *absencing* that allows for *presencing*; the openness that allows for being and that brings things to light. In fact, Heidegger uses the term *Lichtung*, lighting, to illustrate the concept of clearing: 'In the midst of beings as a whole an open place comes to presence. There is a clearing [*Lichtung*]. ... The being can only be as a being, if it stands within, and stands out within, what is illuminated in this clearing'.[8]

Heidegger's clearing is a place in the woods among the trees, which he evokes in *The Thinker as Poet*:

> When the evening light, slanting into
> the woods somewhere, bathes the tree
> trunks in gold. ...
> Singing and thinking are the stems
> neighbour to poetry.

They grow out of Being and reach into
    its truth.
Their relationship makes us think of what
    Hölderlin sings of the trees of the
    woods:
'And to each other they remain unknown,
    So long as they stand, the neighboring
    trunks'.[9]

The canopy of treetops thins, allowing sunlight to stream in, illuminating the white trunks of some trees and casting others in shadow. The forest floor evens out into a small meadow or a flat patch of sun-drenched grass.

Is this what Heidegger imagines when he thinks of clearing? A state-of-being where things and beings are illuminated; truths are un-concealed, but at the same time concealed. The clearing shapes what can appear. What can be, be said or be done is yielded but also withdrawn. Clearing reveals and discloses. If we follow Heidegger's notion that 'human existence is the openness, clearing, or nothingness in which things can manifest themselves',[10] then clearing *is* being. Heidegger writes: 'Only this clearing grants and guarantees passage to those beings that we ourselves are not, and access to the being that we ourselves are'.[11] Heidegger describes the clearing as a kind of nothingness: an 'open center' that 'encircles all that is, like the Nothing which we scarcely know'.[12]

Clearing also cultivates; it gathers space together. It is a bringing forth or a revealing out of emptiness. As such, clearing allows for the new and unpredictable to emerge. It allows for a kind of thinking that is not systematic or logical but rather involves 'rushes of inspiration with potential to surprise the thinker'.[13] To camp in clearing, and to clear space and time for camping, is to flirt with inspiration and surprise. '"Clearing" opens up the possibility for the free play of making place anew'.[14]

In this sense, Heidegger's notion of clearing as the ontological foundation of being in the world and as a condition for creativity is seductively resonant with the ontological experimentality of camping. And yet, it is also somehow incompatible with the actual relations of clearing, camping, and being-together. It takes us only so far. If we apply Heidegger's versions of clearing, *Lichtung*, and nothingness to camping, for example, we lose sight of the way clearing often violently invents emptiness where there was none.[15] For emptiness to exist, the something that must be cleared is effectively excluded. Clearing tilts toward tidiness; it connotes a civilizing impulse in the midst of chaos.

In contrast to Heidegger's connotation of clearing as a demarcated space of being or dwelling, Nuccio Mazzullo and Tim Ingold offer instead the notion of 'being along'.[16] Drawing on their ethnographic fieldwork with Sámi reindeer herders in Finnish Lapland, Mazzullo and Ingold argue that Heidegger's image of clearing may resonate with settlers, whose houses are situated in clearings demarcated by the edge of the woods, but that it 'fails to capture the Sámi sense of what it means to be at home'.[17] For one thing, the Sámi herders live in tents, or *lávvu*, that can be carried, assembled, and disassembled. But this is only part of the reason why Heidegger's fairly static notion of clearing does not hold. Mazzullo and Ingold go on to explain that for Sámi people, home is not about cultivating a 'narrowly circumscribed clearing in the woods', but rather it 'is suspended in movement'.[18] Home is woven into journeys through the woods and along tangles of pathways that knit movement, people, and places together. So we might mobilize Heidegger's notion of clearing by thinking of it not just as being but as being-along. Along these lines, we might also ask about clearing as being-with. In Heidegger's formulation, we also seem to lose sight of each other. Although the being that clearing makes possible is not necessarily atomized in the world, it does appear to be quite solitary in Heidegger's account. Who walks with us on the forest path as the sun illuminates the trunks of the trees? How do my sister and I accompany one another in our violent and creative acts of clearing?

Camping in *chora*

As a somewhat more sociable ontological companion to Heidegger's clearing, we might consider the concept of *chora* to think about camping and campsites as generative in-between places. *Chora* is clearing's cousin.[19] Both concepts gesture toward the emptiness through which *being* becomes possible. But if Heidegger's clearing suggests a somewhat isolated and neutral being-in-the-world, *chora* is more family oriented and more feminine. Although the etymological legacy of 'camp' is most often traced to its Latin origins, it includes variants on the Greek term *chora* as well.[20] The term is used most notably in

Plato's *Timaeus*, where Plato invokes *chora* to refer to the spatial condition for the existence of material objects, a space that functions as an in-between space, 'neither something nor yet nothing', through which the material world is generated.[21] To illustrate, Plato compares *chora* to the wax on which an image is stamped or the odourless oil that carries the scent of a perfume.[22]

In her analysis of Plato's *Timaeus*, Elizabeth Grosz explains that Plato uses the concept of *chora* as a kind of intermediary between being (that which is unchanging) and becoming (that which is subject to change).[23] This is the first paradox Grosz observes in Plato: that he should disrupt his own ontological dichotomy (between being and becoming) with a third concept. She then unravels a second paradox: Plato's claim that *chora*'s 'quality is to be quality-less'.[24] The role of *chora* is to bring being into existence, but it fulfils this role without taking on any qualities itself. Grosz explains:

> It is the space that engenders without possessing, that nurtures without requirements of its own, that receives without giving, and that gives without receiving, a space that evades all characterizations including the disconcerting logic of identity, of hierarchy, of being, the regulation of order.[25]

Plato refuses to imbue *chora* with any specific qualities – to do so would eliminate its generative indeterminacy – and for Plato this seems to go for sexual difference as well. And yet, according to Grosz, Plato explicitly compares *chora* to the role of the female, in particular the 'biological function of gestation'.[26] He associates it with sexually coded terms, describing it in various places as a kind of nurse, matrix, incubator, or womb that 'provid[es] the point of entry, as it were, into material existence'.[27]

Sexual difference is thus both disavowed in and deeply integral to Plato's account of the ontological status of *chora*. What remains, Grosz writes, is a 'disembodied femininity as the ground for the production of a (conceptual and social) universe' that has allowed masculinist modes of thinking to ignore the debt they owe to 'the maternal space from which all subjects emerge'.[28] In other words, the ground on which we camp may be conceptualized as vaguely feminine, but what of the actual women who dwell there?

To puzzle her way through this paradox, Grosz turns to the work of Luce Irigaray, whose feminist reading of *chora* seeks to reappropriate the feminine and maternal dimensions of *chora* and to 'reorient the ways in which spatiality is conceived, lived, and used'.[29] Following Irigaray, Grosz critiques the way phallocentric discourses have used the metaphor of femininity – *chora* – as 'the condition for men's self-representation and cultural production' while simultaneously disenfranchising, disinvesting, and silencing women.[30] This returns Grosz to the question she poses at the beginning of the essay: 'Where and how to live, as whom, and with whom?'[31]

Only now she is asking how we might reconnect the concept of *chora* to the 'female, and especially material, body' and recuperate *chora*, not as a space that contains or obliterates women but as a '*viable* space and time for women to inhabit as women'?[32] The answer is somewhat nebulous. Grosz concludes her essay by calling on readers to 'experiment with and produce the possibility of occupying, dwelling or living in new spaces, which in their turn help generate new perspectives, new bodies, new ways of inhabiting'.[33] The sentiment is inspiring, but what exactly might such an experiment look like? How do we live together, and yet live with difference – including sexual difference – in these new spaces? More to the point, can *chora* disrupt masculinist spaces without disavowing sexual difference anew?

Grosz's call to reappropriate *chora* has resonated in critical feminist approaches to tourism and hospitality, perhaps most famously in Betsy Wearing's book *Leisure and Feminist Theory*. Here, Wearing uses the concept of *chora* to interrupt the phallocentric models of place that often underpin, for example, male-centred urban planning. To do this, Wearing emphasizes two aspects of *chora*: that it is *inhabited* space, 'a space whose meanings can be constantly redefined by its inhabitants'; and that it is *interactive* space, one that 'involves interaction with others and ... becomes integrated into one's sense of self'.[34] Through the concept of *chora*, we might imagine the camp as an experiment in living with difference. As Wearing puts it, *chora* is a space for 'alternative subjectivities', and also perhaps for alternative relations between and among people (hospitality) and alternative life politics as we practise the art of living together (conviviality)[35] and living near one another (neighbouring).[36] In Wearing's analysis, *chora* is 'a safe space for social interaction which women as well as men may enjoy and which may enhance rather than diminish a sense of self or identity'.[37]

Wearing's idea about alternative identities is echoed in Hailey's suggestions that 'camp space, as a mobile family space or, more broadly, as a community space, accommodates another kind of identity'.[38] In this sense, the concept of *chora*, along with clearing, brings our attention to the generative potential of empty space and unfinished architectures of tourism – both concrete and metaphorical. Emptiness is necessary in order to be, to become, to make the next move. In other words, *chora* contains the 'possibility of stretching toward the future, of becoming'.[39]

What new kinds of relations and subjectivities are at stake? Wearing starts by telling us what they will *not* look like. She presents the typical mode of consuming these spaces, which involves, not coincidentally, the typically male version of the disembodied tourist gaze: the *flâneur*. Wearing sees *flâneurs* not as 'interacting people' but as standing back, observing the world through fleeting observations and from an objective distance.[40] In contrast, and drawing on the idea of *chora* as a space that allows women to belong and to become rather than merely to be gazed upon, Wearing introduces the figure of the *choraster*. Unlike the *flâneur*, the *choraster* is 'a feminized conceptualization of those who interact in a constructive or creative way with others in city leisure spaces'.[41] *Chorasters* inhabit and interact with the space, and with each other. In contrast to the detached masculine voyeur that is the *flâneur*, Wearing's feminine *choraster* moving through shared public leisure space comes across as a socially and spatially embedded subject.

Pitching the tent

Our next camping expedition takes place in an episode of the popular American television sitcom, *Parks and Recreation*, which aired in 2011. The episode, titled 'Camping',[42] centres on Leslie Knope, an overachieving, mid-level bureaucrat in the Parks and Recreation Department of the small city of Pawnee, Indiana, who is under pressure to come up with a creative fundraising idea for the town. Usually prolific, Leslie finds herself tapped out, unable to generate any ideas. She decides to pack her staff into the department's van and take them on a camping trip to jumpstart their creativity. Once the group has set up camp, several story lines unfold, two of which I want to focus on here.

First, despite leaving the office and going into 'the dark woods where ideas are born',[43] Leslie and her staff fail to think of any new ideas. The problem, we soon discover, is not with camping itself, but rather with Leslie's unceasing effort to be creative. As it becomes clear that no good ideas are forthcoming, Leslie resolves that they will stay up all night working until they come up with something great. She bangs on a metal pan to keep everyone awake, insisting that the path to an outcome is work, work, and more hard work. One of Leslie's colleagues complains that brainstorming is a waste of time; it doesn't matter what they come up with since Leslie's idea will be so much better. And Tom, Leslie's assistant, concedes: 'The truth is she's better at this than we are by like a thousand miles'. Instead of inspiring creativity, Leslie's over-functioning has made her staff complacent. In this case, more – more work, more thinking, more ideas – is less.

Meanwhile, a second story line with a parallel theme unfolds around Tom's tent. Dubbed 'Thunderdome' by Tom, the tent is a cushy materialization of modern consumer culture. A brief tour of Tom's tent reveals a full-size bed, a wool carpet, a flat screen television, an X-Box video game console, a digital video recorder, incandescent lights, a soft-serve ice cream machine, an electric head massager, a miniature pool table, a fondue set, and a dog bed repurposed as a sofa – all ordered out of the Sky Mall catalogue. Frequent fliers in particular will be familiar with this catalogue, a ubiquitous feature of every airplane seatback, as a paean to the manufacture – and instant satisfaction – of dozens of consumer needs you never knew you had.

Tom powers his consumer appliance fantasy by wiring his tent to the van's battery. As the last of the battery power drains away and Tom's tent goes dark, we hear him cry: 'No! I was TiVo'ing *Cupcake Wars*! Look, I'm used to a certain level of comfort in my life and I didn't want to sacrifice that!' Tom and his tent full of consumer goods defy the ethos of discomfort and emptiness that gives camping its moral value. This overabundance of stuff leaves no room to move. With a dead battery in the van and no other transportation, the group is stuck in the woods. Tom's unwillingness to sacrifice his own comfort uses up all the power and literally immobilizes the group.

Breaking camp

Rolling up the sleeping bags. Pulling down the tents. Packing an inhabited world back into our rucksacks or into the back of the car. Clearing out the clearing in which we have, momentarily, both shifted and reproduced the very grounds of everyday life. Camping reminds us that we live in an unfinished world. When we consider camping through the ontological prisms of clearing and *chora*, we can begin to open up to these potentialities. Hailey observes that 'camp constructions remain in an undeveloped, unfinished, and incomplete state regardless of their apparent degree of permanence. The constructions of *camping* are not things made but are things being made, or more precisely things becoming'.[44] There is something about camping that conjures up a self-conscious awareness of the unfinalized condition of our social worlds. At home, this contingency drifts out of sight, replaced in our everyday consciousness by the weighty and ongoing there-ness of walls and chairs, beds and dishes, bodies and clothes, routines and habits.

After all, this is what a home does – it hold us together as if we were already coherent beings to begin with. Life becomes consolidated through fixed domestic structures, interior arrangements, and repetitive routines that feel as though they were always already there, making it easy to forget that the spatial and social arrangements of our everyday lives are also always in process.

This is not to say that home is not also a contingently reproduced site of social life, but as with other routinized spaces, this contingency has a tendency to fade into the background.

Just because we forget, however, does not mean it ceases to be. We are always making and remaking the textures and interactions that come to constitute our 'day' and our 'self' and our 'family' and our 'home'. The seeming endurance – especially in Western lifestyles – of things like living rooms and kitchens and bodies and families belie the incessant labour of inhabiting life. In his meditations on home in *The Poetics of Space*, Gaston Bachelard writes: 'The house thrusts aside contingencies, its councils of continuity are unceasing. Without it, man would be a dispersed being. It maintains him through the storms of the heavens and through those of life'.[45]

Camping tosses us back into those storms. It reminds us of the *evitability* of the way things are. Our efforts at cobbling together a home-away-from-home – of clearing a site and pitching a tent – bring us to the exhilarating (and terrifying) realization of the could-be-otherwise-ness of what has come to seem so solid. In camping, our social arrangements and material worlds are exposed for the tenuous filaments and contingent patterns they are. Their fundamentally unfinished nature is laid bare.

Unfinishedness is a theme that emerges often in the arts, architecture and pedagogy but rarely in the context of tourism. Artists have long played with gaps and silences to generate creative possibilities. Think of improvisational jazz, for example, or the ontologies of silence that Soile Veijola explores in her chapter on silent communities. In architecture, unfinishedness is about the generative interplay between absence and presence. In the 1970s, architect Lars Lerup advocated for 'building the unfinished'.[46] What Lerup meant was not that the construction work should be abandoned before the building is finished, but rather that unfinishedness is built into the design from the start. Lerup argued that human action was 'a complicated matrix with unknown combinations, the result of which is considerable unpredictability, a marvelous unfinishedness and openness' and worried that if this fact was ignored, architecture would become 'simply utilitarian' and 'design [would] become dull, repetitive and mechanical'.[47] This is why Lerup insisted that we must become comfortable with imprecision and unfinishedness if we hoped to leave room for unpredictable future interactions and configurations beyond the view of our current horizon.[48] What is left unfinished is not a void, but perhaps something more like a clearing or *chora* – an interstitial space that allows for becoming.

This is what the radical philosopher and educator Paulo Freire is getting at when he writes about unfinishedness. He sees it as an essential quality of the human condition in the sense that 'men and women [are] beings in the process of *becoming* ... unfinished, incomplete beings in and with a likewise unfinished reality'.[49] Our unfinishedness is what makes it possible for us to learn, to grow, to become. For Freire, our awareness of our own unfinishedness, and our concomitant quest for completeness, also drives our curiosity and makes humans eternal seekers. We are 'destined by our very incompleteness to seek completeness, to have a "tomorrow" that adds to our "today"'.[50] For this reason, Freire presents unfinishedness as a source of hope, pushing us toward new horizons: 'wherever there are men and women, there is always and inevitably something to be done, to be completed, to be taught, and to be learned ... to become "more"'.[51]

Freire urges us to embrace unfinishedness as the basis of becoming, but unfinishedness is a slippery concept to add to our ontological toolkit for tourism. Its implicit associations with 'completion' and with 'more' seem to set unfinishedness on a teleological trajectory that defies the indeterminacy of becoming. Then again, perhaps this aporetic tension between unfinishedness and completion – between impulse and impossibility – is what holds the door of becoming open. As Hailey reminds us, 'unfinished' does not necessarily 'provide valuation or quantification of completeness or of a lack. The unfinished is not necessarily the incomplete'.[52] In tourism, however, we often turn unfinishedness away as an unwelcomed guest. If those glossy magazine pages and Tom's tent are any indication, the imaginaries (if not the realities) of tourism do not bend toward unfinished, but rather tend toward the plumply replete and toward catering to, fulfilling, and satisfying every possible desire. Can unfinishedness disrupt this tourism fantasy?

On the horizon

Our final campsite is in the Danish countryside outside of Copenhagen. It is late May of 2009, and some of the world's most talented chefs are tramping through the woods gathering shoots and mushrooms, pulling up fistfuls of wild garlic, and fishing in the river. Then, using as few mechanized appliances as possible, they prepare these ingredients for dinner. This is the launch of a culinary experiment called *Cook It Raw*. The event brought together a dozen renowned chefs[53] and tasked them with a specific challenge: to create cutting edge cuisine with an eye toward environmental apocalypse. On the premise that even in a future scenario of food and energy scarcity we should still be able to eat fantastically well, the chefs were charged with using energy-saving techniques to prepare native and wild food foraged from the land.

The experiment – which has since been repeated in Italy, Lapland, Japan and Poland – has been depicted by television chef Anthony Bourdain as a 'summer camp for culinary hot shots'.[54] In fact, several journalists and some of the participating chefs themselves refer to *Cook It Raw* as camping.[55] René Redzepi, famous as the foraging head chef of the world-renowned Noma restaurant in Copenhagen, describes the event like this:

> It felt like going to camp: you get on the bus and it's really excit-ing because you are free from your everyday responsibilities. *Cook It Raw* often feels like free-falling into the unknown. You have to step out of your daily routine and put yourself on the line. In a restaurant, you have a perfect setup, everything functions as you want it. Then, for the duration of *Cook It Raw*, you erase that. You have to do things very differently, and maybe it will fuck up, but it is this change of routine that pushes you into new ideas and new ways of looking at things.[56]

What Redzepi calls on here is the interplay between camping as a kind of clearing – freedom from the sediment of everyday responsibilities – and as the unknown, the unformed, the unperfected. The rules by which the chefs must live and cook – foraging only local ingredients and preparing them in an energy-sustainable way – are boundaries that promote rather than inhibit creativity. The emphasis is on gathering; on pulling something together out of nothing.

Camping offers a discourse that helps us make sense of this par-ticular experiment with food, nature, place, and technique. Here, the ethos that 'less is more' becomes a strategy for living in a likely future. The limits imposed on the chefs are seen as a source of crea-tivity, 'sparking new personal and creative connections' and pushing the chefs to become more 'flexible thinkers'.[57] We learn from the *Cook It Raw* website:

> Creativity is a hungry, restless beast. It rambles in the forest crav-ing fresh knowledge, waiting to be fed and quenched. *Cook it Raw* gatherings satisfy this hunger, pushing chefs out of their comfort zones and into the dark woods where ideas are born.[58]

On the final night of the *Cook It Raw* event in Japan in 2011, the chefs turned out what one journalist described as 'high-concept for-est floor tableaux'[59] with concoctions named 'Frustrated Mackeral', 'Earth Marrow', and 'Norwegian Wood'. Here, the chefs do not just clear the forest floor; they gather it onto their plates. The organizers of *Cook It Raw* describe the experiment as an exercise in 'controlled improvisation', though perhaps we can also understand it in Heideggerean terms with boundaries as starting points, clearing as gathering and the forest floor as a source of nourishment.

Cook It Raw is not about camping memories but about camping as the future condition of human existence after environmental and social collapse. According to the *Cook It Raw* website, we need not wait for the new necessities of disaster to prompt a sense of innovation. By learning how to prepare basic proteins and forage edible flora, we can begin now to adapt to our hot new world:

> When the environmental apocalypse is close, creativity will save the world. What if in 2050 jellyfishes and giant squids will be the only creatures left in the black, polluted sea? ... What if foraging will soon be your last resort? Environmental issues, like global warning, might not shine lively bright in the public eye. Still, they are real and evolving (read 'getting worse') behind the scenes. ... Now that the damage is done, it's time to think about adaptive strategies, starting right from what is in your plate. ... Local and seasonal raw materials, energy-saving cooking techniques, and a hint of creativity could be appropriate ingredients of the answer, on the way out of the greenhouse-effect nightmare.[60]

Indeed, it is the lessons we learn from camping – lessons about doing without, making do, and making anew – that will serve us well in this future world. The website continues:

> It is clear that nothing in gastronomy discourse could be as abruptly innovative as this 'high-cuisine-in-a-campsite' experiment. Thus, there are good reasons to foresee a happy, lacking ending: less is more if creativity is at work.[61]

This happy, lacking future is always impending. It disabuses us of the fantasy of completion that unfinishedness might imply. Even though it makes its presence felt today, in the foraging and culinary skills displayed by the chefs, it never actually arrives. It lingers perpetually on the horizon.

Conclusion

Clearing, *chora* and unfinishedness, as spatial ontologies of camping, can open up different ways of thinking about the pleasures of tourism – not as an escape or a retreat (a retreating) in the sense of a passive disavowal of our problems or abdication of responsibility but as a withdrawal (absencing) that opens up space for altogether new conditions and ways of being to emerge. These ontologies leave room for serendipity, disruption, unpredictability, uniqueness, arbitrariness and *being*.[62]

Consider, for example, the tourism architecture *par excellence*: the hotel room. The designer's skill reflected in the hotel is the ability to anticipate the guest's every need before it is spoken or even a conscious thought. The light switches, soap, waste basket, water glasses, towels, toilet paper, curtain pulls, pillows, hangers, television remote, room service menu are all at hand. Ideally, they materialize just as that hand reaches out for them. But what neither the conventional hotel room nor the all-inclusive resort can anticipate, and therefore what they cannot provide – for it cannot be provided, as such – is the open space for alternative ways of living in that place. Indeed, there is only one (ideal, normative) way to live in a hotel room. (Other ways may be possible, but become transgressive uses of the space that serve to normalize the status quo.)

It is no accident that the place where tourism studies meets Agamben's paradigm of the camp is in the hotel, and more precisely in the all-inclusive tourist resorts of Ibiza and Club Med.[63] These meeting grounds may seem surprising and even violent, given the vast chasm between the horror of the concentration camp and the hedonism of the tourist enclave, but the parallels are not coincidental. As Claudio Minca points out, 'both institutions are based on a radical extraterritorial status and are made possible by *a regime of exception* – and sometimes by *the very suspension of the norm*'.[64] Agamben defines the camp as 'the space that is opened when the state of exception begins to become the rule'.[65] According to Bülent Diken and Carsten Bagge Laustsen, this means that the camp may take the spatial form of extreme deprivation, such as concentration camps or refugee camps or favelas, but it may also take the form of extreme privilege in 'benevolent' camps, such as tourist resorts, that 'repeat the logic of the exception', only this time for the 'winners'.[66] And sometimes the very same space takes both of these forms simultaneously, such as when slums and favelas become the object of the tourist gaze and shanty towns become all-inclusive resorts, or when hotels become prisons for asylum seekers and prisons become hotels for experience-hungry tourists.[67] Agamben explicitly extends the paradigm of the camp to the seemingly innocuous space of the hotel, where, as in concentration camps or refugee camps, 'the normal order is *de facto* suspended'.[68] Maybe this is why we are surprised and at the same time not at all surprised when hotels become prisons become hotels.

The all-inclusive tourist resort – epitomized by Club Med in Diken and Laustsen's account – thus figures as an example of the 'nomos' of the camp: 'the tourist enters an enclosed, exceptional and "duty-free" zone, "taken outside" home, everyday routine and familiar social/ moral contexts'.[69] Tourists are urged to go barefoot, strip down to bathing trunks and, as the marketing brochures put it, leave 'civilized' life behind.[70] In this Club Med enclave, the rules of social order are suspended, the constraints of places are inverted, and the usual determinants of identity and obligation – and even clothes – are shrugged off. The festival ambience of the holiday resort offers 'an opportunity to become naked, that is, to get rid of one's markers of identity'.[71]

In a fitting paradox, these most carnivalesque atmospheres are generated in some of the most strictly classed and rigidly choreographed spaces that tourism has to offer. Given this premise, the critique that follows is fairly easy to predict. Playing the paradox out to its logical conclusion, Diken and Laustsen argue that the hedonistic excesses in these zones of exception may appear to disturb the social order, but ultimately the tourist's efforts to transcend the rules result only in reasserting them.[72] This conclusion aligns with the paradox of the camp, as Agamben explains it:

> The camp is a piece of land placed outside the normal juridical order, but it is nevertheless not simply an external space. What is excluded in the camp is, according to the etymological sense of the term 'exception' (ex-capere), taken outside, included through its own exclusion. But what is first of all taken into the juridical order is the state of exception itself.[73]

In other words, the camp stabilizes the status quo to which it is an exception.[74]

Given that the camp in question – the all-inclusive Club Med resort – is something of a caricature, we should not be too surprised to bump up against this ontological dead-end. From this distance, it is practically impossible to tell the *flâneurs* from the *chorasters*, as we will see in Veijola's chapter in this book. For now, however, perhaps Diken and Laustsen are asking too much of the camp and its campers by expecting them to radically transform the current arrangements of social, political and cultural life. In *The Coming Community*, Agamben paints a far more modest picture of global transformation. Despite his bleak diagnosis for this world, Agamben suggests that redemption will not involve a 'momentous transformation', but rather 'a "small displacement" that will nonetheless make all the difference'.[75] In the world to come, he writes, 'everything will be as it is now, just a little different'.[76] It is along these lines that I have tried to understand camping, not as a radical transgression of the social order but rather as a disruption that reminds us of the small ways in which life can be lived differently. Camping allows for the ruptures and intervals in which something else could happen. It reminds us that the world as-it-is is loosely stitched together, and that it could be otherwise.

***

By way of contrast to the episode of arriving at the hotel or the all-inclusive resort where everything, including the bed, is made (by the maid) for us, the campsites I have described in this chapter offer few amenities in anticipation. (As promised at the start, I have tried to leave my camp story incomplete, unfinished.) But they do offer some. Perhaps *just enough*, which is what makes the camp not the opposite of the hotel room but a special configuration in comparison with the hotel. Think of camping in nature. The clearing – in which to pitch the tent – and the trees, meadows, streams, hills and mountains, and a few built facilities afford certain embodied and social performances,[77] like using a tree branch as a hook for hanging a bucket or a towel. But camping does not anticipate the tourist's every need. To do so would be to pull the plug on possibility, as we saw in the episode with Tom's tent.

The not-at-hand-ness of amenities is what prompts creative, serendipitous, alternative actions. The tourist – like the chef foraging in the forest – becomes a *bricoleur*, creatively re-appropriating objects, spaces, openings where they can be found for the desires and needs of the moment. (For more on such serendipities, see the chapter by Alexander Grit.) We might even say that in its exquisite anticipation, the hotel creates the very needs and desires it fulfils, as a way of controlling or mitigating the excesses of need and desire. This is the tyranny of luxury. The kind of camping I have been discussing, on the other hand, clears a space for unpredictable needs and desires, performances and possibilities, to emerge and to be worked through. Through the prisms of clearing and *chora*, camping brings us back around to a familiar ontological story: our social worlds are always in process, never finalized, and therefore always open to becoming arranged and lived *differently*. Camping allows us to push past our known horizons toward new ways of doing, being, and living together.

Notes

1. Steven Shaviro, *Without Criteria: Kant, Whitehead, Deleuze, and Aesthetics*, Cambridge, MA, MIT Press, 2009, p. 146. I am grateful to Olly Pyyhtinen for pointing me toward this theoretical approach.
2. Ibid.
3. Charlie Hailey, *Camps: A Guide to 21st-Century Space*, Cambridge, MA, MIT Press, 2009, p. 3. RV culture refers to the mobile camping culture associated with recreational vehicles, or caravans. This mode of camping is particularly common in the continental United States, Australia and parts of Europe.
4. For additional examples, see Fabian Frenzel, 'Exit the System: Crafting the Place of Protest Camps Between Antagonism and Exception', Working Paper, University of the West of England, 2011, pp. 1–36; Tim Cresswell, *In Place Out of Place: Geography, Ideology, and Transgression*, Minneapolis, University of Minnesota Press, 1996; Orvar Löfgren, *On Holiday: A History of Vacationing*, Berkeley, University of California Press, 1999.
5. Charlie Hailey, *Campsite: Architectures of Duration and Place*, Baton Rouge, Louisiana State University Press, 2008.
6. Giorgio Agamben, *Homo Sacer: Sovereign Power and Bare Life*, translated by D. Heller-Roazen, Stanford, Stanford University Press, 1998.
7. Avtar Brah, *Cartographies of Diaspora: Contesting Identities*, London, Routledge, 1996.
8. Martin Heidegger, *Poetry, Language, Thought*, translated by A. Hofstadter, New York, HarperCollins, 1971/2001, p. 51.
9. Ibid., p. 14.
10. Michael E. Zimmerman, 'Heidegger, Buddhism and Deep Ecology', in Charles B. Guignon (ed.), *The Cambridge Companion to Heidegger*, Cambridge, Cambridge University Press, 1993, p. 241.
11. Heidegger, *Poetry, Language, Thought*, p. 51.
12. Ibid.

13. Adam Sharr, *Heidegger for Architects*, London and New York, Routledge, 2007, p. 11.
14. Hailey, *Campsite*, p. 78.
15. Thank you to Olli Pyyhtinen for this thought.
16. Nuccio Mazzullo and Tim Ingold, 'Being Along: Place, Time and Movement Among Sámi People', in Juergen Ole Baerenholdt and Brynhild Granås (eds), *Mobility and Place: Enacting Northern European Peripheries*, Aldershot, Ashgate, 2008, pp. 27–38.
17. Ibid., p. 30.
18. Ibid., p. 35.
19. See Joanne Faulkner, 'Amnesia at the Beginning of Time: Irigaray's Reading of Heidegger in the Forgetting of Air', *Contretemps*, 2, May 2001, pp. 124–141.
20. Hailey, *Campsite*, p. 55.
21. Elizabeth Grosz, 'Women, *Chora*, Dwelling', in E. Grosz (ed.), *Space, Time and Perversion*, London and New York, Routledge, 1995; and see Inger Birkeland, *Making Place, Making Self: Travel, Subjectivity and Sexual Difference*, Aldershot, Ashgate, 2005, p. 115.
22. Tom Boellstorff, 'Placing the Virtual Body: Avatar, Chora, Cypherg', in Frances E. Mascia-Lees (ed.), *A Companion to the Anthropology of the Body and Embodiment*, Malden, Blackwell, 2011, pp. 504–520.
23. Grosz,'Women, *Chora*, Dwelling'.
24. Ibid., p. 113.
25. Ibid., p. 116.
26. Ibid., p. 115.
27. Ibid, p. 114. I am grateful to Soile Veijola for helping me emphasize the significance of sexual difference in these discourses.
28. Ibid., pp. 113, 121.
29. Ibid., p. 120.
30. Ibid., p. 124.
31. Ibid., p. 113.
32. Ibid., pp. 120, 124.
33. Ibid., p. 124.
34. Betsy Wearing, *Leisure and Feminist Theory*, London, Sage, 1998, p. 133.
35. For more on hospitality as conviviality, see Jennie Germann Molz, *Travel Connections: Tourism, Technology and Togetherness in a Mobile World*, London and New York, Routledge, 2012, Chapter 5.
36. Soile Veijola and Petra Falin, 'Mobile Neighbouring', *Mobilities*, online first, doi 10.1080/17450101.2014.936715.
37. Wearing, *Leisure and Feminist Theory*, p. 127.
38. Hailey, *Campsite*, p. 107.
39. Wearing, *Leisure and Feminist Theory*, p. 137.
40. Ibid., p. 133.
41. Ibid.
42. 'Camping', *Parks and Recreation* (television program), NBC, 24 March 2011, Season 3, Episode 8.
43. Cook It Raw,www.cookitraw.org, website accessed on 28 April 2013.

44. Hailey, *Campsite*, p. 7.
45. Gaston Bachelard, *The Poetics of Space*, Boston, Beacon Press, 1958/1994, p. 7.
46. Lars Lerup, *Building the Unfinished: Architecture and Human Action*, London, Sage, 1977.
47. Lerup, cited in Jeremy Till, *Architecture Depends*, Cambridge, MA, MIT Press, 2009, p. 107.
48. Ibid.
49. Paulo Freire, *Pedagogy of the Oppressed*, New York, Continuum, 1970/2000, p. 84.
50. Paulo Freire, *Pedagogy of Freedom: Ethics, Democracy, and Civic Courage*, Lanham, MD, Rowman & Littlefield, 1998, p. 79.
51. Ibid.
52. Hailey, *Campsite*, pp. 31–32.
53. What should we make of the fact that all but one of the *Cook It Raw* chefs are men? Is this a harbinger of the 'hostessing society' that Soile Veijola and Eeva Jokinen (2008) see coming into view and into practice in 'new work' that requires both men and women to inhabit the skills of the hostess? Perhaps, but I think other explanations are more likely. To paraphrase Jamaica Kincaid (*A Small Place*, 1988), there is a world of something in this that I can't go into right now.
54. Anthony Bourdain, 'Japan: Cook it Raw', *No Reservations* (television program), Travel Channel, 7 May 2012, Season 8, Episode 5.
55. *Phaidon*, '16 of the World's Most Intrepid Chefs Camping Out in the Wilderness of Lapland for a Week', http://uk.phaidon.com/edit/food/articles/2010/september/28/16-chefs-go-to-lapland/, website accessed on 30 April 2013.
56. Tim Lewis and Marie-Claude Lortie, 'Cook It Raw: For Chefs, It's Like Free-Falling into the Unknown', *The Observer*, Saturday 20 April 2013, http://www.guardian.co.uk/lifeandstyle/2013/apr/21/cook-it-raw-eight-chefs-recall, website accessed on 30 April 2013.
57. Cook It Raw, www.cookitraw.org.
58. Ibid.
59. Elizabeth Gunnison, 'How to Eat Ramen Like Anthony Bourdain', *Esquire*, http://www.esquire.com/blogs/food-for-men/anthony-bourdain-ramen-upgrades-8682344, website accessed on 30 April 2013.
60. Cook It Raw, www.cookitraw.org.
61. Ibid.
62. Rosalyn Diprose, 'Building and Belonging Amid the Plight of Dwelling', *Angelaki: Journal of Theoretical Humanities*, 16.4, 2011, pp. 59–72.
63. Claudio Minca, 'The Island: Work, Tourism and the Biopolitical', *Tourist Studies*, 9.2, 2010, pp. 88–108; Bülent Diken and Carsten Bagge Laustsen, *The Culture of Exception: Sociology Facing the Camp*, London and New York, Routledge, 2005; and see Frenzel, 'Exit the System', p. 2.
64. Minca, 'The Island', p. 102, italics in original.
65. Agamben, *Homo Sacer*, p. 96.

66. Diken and Laustsen, *The Culture of Exception*, p. 9.
67. On favela tourism in South America and slum tourism in South Asia, respectively, see Bianca Freire-Medeiros, 'The Favela and its Touristic Transits', *Geoforum*, 40.4, 2009, pp. 580–588; and Anya Diekmann and Kevin Hannam, 'Touristic Mobilities in India's Slum Space', *Annals of Tourism Research*, 39.3, 2012, pp. 1315–1336; as well as the 2012 special issue on slum tourism in *Tourism Geographies*, 14.2, edited by Fabian Frenzel and Ko Koens. For an example of shanty towns offered as all-inclusive tourist retreats, see http://www.emoya.co.za/p23/accommodation/shanty-town-for-a-unique-accommodation-experience-in-bloemfontein.html, website accessed on 30 December 2013. On hotels repurposed to house asylum seekers, see Sarah Gibson, 'Accommodating Strangers: British Hospitality and the Asylum Hotel Debate', *Journal for Cultural Research*, 7.4, 2003, pp. 367–386. Examples of prisons repurposed as hotels include the Arlamow Hotel, where Lech Walesa, leader of Poland's Solidarity movement, was detained in the 1980s. And the *Huffington Post* reports on five prisons that have been transformed into luxury hotels: http://www.huffingtonpost.com/justluxe/5-prisons-turned-into-luxury_b_2885226.html, website accessed on 15 December 2013.
68. Agamben, *Homo Sacer*, p. 99; and see Ramona Lenz, '"Hotel Royal" and Other Spaces of Hospitality: Tourists and Migrants in the Mediterranean', in Julie Scott and Tom Selwyn (eds), *Thinking Through Tourism*, Oxford, Berg, 2010, pp. 209–230.
69. Diken and Laustsen, *The Culture of Exception*, p. 112. (Their description draws on Marc Augé's concept of 'non-places', presented in *Non-places: An Introduction to Supermodernity*, London, Verso, 1995.)
70. Ibid.
71. Ibid., p. 116.
72. Ibid.
73. Agamben, *Homo Sacer*, pp. 96–97.
74. Frenzel, 'Exit the System', p. 2.
75. Sergei Pozorov, 'Why Giorgio Agamben is an Optimist', *Philosophy Social Criticism*, 36, 2010, pp. 1053–1073.
76. Giorgio Agamben, *The Coming Community*, translated by M. Hardt, Minneapolis, University of Minnesota Press, 1990/2007, p. 57. And see Jessica Whyte, '"A New Use of the Self": Giorgio Agamben on the Coming Community', *Theory and Event*, 13.1, 2010, pp. 1–19.
77. See James J. Gibson, *The Ecological Approach to Visual Perception*, Boston, Houghton Mifflin, 1979; Eeva Jokinen and Soile Veijola, 'Mountains and Landscapes: Towards Embodied Visualities', in David Crouch and Nina Lübbren (eds), *Visual Culture and Tourism*, Oxford, Berg, 2003, pp. 259–279.

3
Paradise with/out Parasites

Olli Pyyhtinen

Introduction

In this chapter, the untidy guest appears in the disguise of a *parasite*.[1] (However, I have barely managed to utter these words, when the critical, all-knowing sociologist – swiftly and almost immaterially – glides to the scene and immediately objects:)

– Wait a minute! What are you talking about? A parasite?

– Yes, I reply, taken by surprise by this intrusion into my narrative privacy. – The parasite is something or someone who benefits at the expense of the host. The parasite always takes, never gives.

– But aren't parasites by necessity always small invertebrates? he demurs, and continues:

– Tapeworms, fleas, vermin, flukes, lice and the likes, but surely no humans? If you are telling me that a tapeworm has ever come and knocked on your door or been invited to the *table d'hôte*, then I must say that you've lost it. I thought you were a proper, respectable Sociologist. You have got to be kidding. Besides, all this makes my skin itch...

– Let me assure you that most of the parasites I am referring to will not bite you. Yet please bear in mind that even animal parasitism is all about guests and hosts. As philosopher Michel Serres suggests in his book *The Parasite*, '[t]he animal-host offers a meal from the larder or from his own flesh; as a hotel or a hostel, he provides a place to sleep, quite graciously, of course'.[2] Serres proposes that the language of the science called parasitology 'bears several traces of anthropomorphism'.[3] It 'uses the vocabulary of the host:

hostility or hospitality',[4] and thus its understanding of parasitic relations is to a great extent shaped by our sense of ancient customs and
habits related to hospitality, table manners, hostelry and relations
with strangers.

I seem to capture my audience, so I go on:

– And just think about humans. Aren't we in fact universal parasites in the sense that everything and everyone around us is a hospitable space? Just think about it: animals and plants are our hosts, and
we their guests. We eat, milk and use them in all sorts of ways. What
is more, we do not stop there, to beings and life forms we consider
inferior to us. We even want to parasite our own kind, our fellow
humans, just as much as they wish to enjoy our own hospitality.[5]

The critical sociologist intervenes:

– So the human being is a louse to another human being? *Homo
homini pediculus*? It's a rather gloomy picture that you paint of
humankind. But the brush you use is missing out something essential. Is it not so that parasites always live *within* their hosts? Humans
surely do not live inside other animals, even if they exploited them!

I continue:

– In fact, don't you know that *The Onion* once reported of a man
living inside Nicholas Cage?[6] Of course, that was only a joke, but
don't we live within the body of a host when we wear a Norwegian
knit or shoes made of leather, for instance? And mind you, we eat,
kick and sleep the first nine months of our lives within the belly of
our mother. However, that is beside my point.

I lean towards my critical, uninvited companion:

– Only some types of parasitism, not all, come down to living
within the host. To parasite is, essentially, to eat *next to* the host. The
parasite lives *on*, *with*, *beside* and *by* one's host. The etymology of the
word 'parasite' is informative here. In it, as Serres notes, the 'prefix
para- means "near", "next to", measures a distance. The *sitos* is the
food'.[7] So in that regard parasites coming in human shape and living
off others are not out of the question. After all, the parasite is, literally, an *entre-preneur*, someone who occupies the position in-between,
as the French word conveniently expresses it. The parasite is always
in-between, in the position of the third.

– Ha-ha, my interlocutor remarks, unamused. – Are you trying to
talk your way out of my allegations by making corny Marxist jokes
about capitalists?

– Oh no, I am being dead serious. And I think that, on the one hand, you are right that the parasitic logic presents a rather pessimistic image of human relations. Many of the relations that we tend to regard as being symbiotic are in fact parasitic – that is, abusive instead of being based on balance and equitable exchange. However, on the other hand, the idea of parasitism suggests that our relations of abuse and exploitation are less violent than we perhaps tend to believe. We are not predators. The host is not prey, but there is generosity and hospitality involved. The parasite always needs a hospitable space around it. The host gives and the parasite takes. The parasite's very life and existence depend on hospitality; when hospitality and giving come to an end, parasitism ends too.

– So that's all there is to it? snorts my companion. – There is nothing 'new' or 'original' about this! To me it seems that the notion of the parasite does nothing more than merely restate the old problem of the 'free rider', albeit in a slightly different and admittedly more obscure vocabulary.

I'm beginning to feel slightly irritated by my interlocutor's belligerence by now. However, I manage to hold my temper and defend myself:

– The notion of the parasite is irreducible to that of the free rider. In French, there are three meanings to the word parasite: (1) in its biological sense, a parasite is an organism feeding on another one without benefiting its host in any way; (2) in the anthropological sense, an abusive guest, who takes without giving anything in return (indeed, unlike the biological parasite, the social parasite does not necessarily live *in* its host, but just *by* it); (3) and in information theory, it designates noise, static, a break in the message. Please remember that the neighbouring function of eating is making noise: the open mouth that eats also emits sound.

– Hmm, the critical sociologist murmurs. – Sounds awfully messy. I can vaguely see the connection between the first two senses, but what on earth could they possibly have in common with the third one?

– That is the tricky part. There is in fact no immediate connection between them, but they only share a similarity of form, an isomorphism. Each of the three meanings displays a relation of a similar kind: *a simple, irreversible arrow.*[8] The parasite is the one who or that which *intervenes* and *interrupts*.

I take a deep breath and continue in a conversational tone:

– May I tell you a story that illustrates this nicely and displays all the aforementioned meanings? Serres's *The Parasite* begins with a recapitulation of a fable by La Fontaine of the city rat who has invited the country rat for a visit. The rats chew and gnaw their meal with absolute delight on a Persian rug. The meal consists of nothing but scraps, bits and leftovers, but for the country rat, at least, it makes a royal feast, since in the country they only eat soup. However, the feast is cut short, as the rats hear noise from the door. The noise made by their cutting and nibbling has woken up the head of the house, the tax farmer, who has now got out of bed to determine the origin of the disturbing sounds. According to Serres, all relations in the story are parasitic by nature. The country rat, though being an animal, is a parasite in the anthropological sense: a guest at a banquet, exploiting its host, the city rat, and thus intervening between the latter and its meal. The city rat, living in the tax farmer's house and feeding on his leftovers, is a parasite in the biological sense: the city rat taxes the tax collector. But the tax farmer, too, is a parasite. He has produced nothing in his own right, neither cheese nor ham nor oil but, by the powers of his position and the law, he profits only from the work of the peasant, himself a parasite of the earth and its fruits. And, finally, the noise that the rats make is parasitic in the third sense of the term, namely that employed in cybernetics, as it wakes up the tax farmer from his sleep, and he in turn becomes a parasite interrupting the feast. One parasite chases another out one after the other. The parasite is 'a noise of the system that can only be supplanted by a noise'.[9]

– One-way relations? exclaims the critical sociologist. – I can't accept that. All relations involve reciprocity! And let me remind you, you yourself have said so in your book on Simmel.[10] Let me see, on page 78, for example, you write: 'Interactivity forms the ontology of the world'. So there you have it! Check and mate! I could also cite dozens of other pages where you suggest that all our relations are based on exchange.

– Please forgive me for the sins of my youth! I beg, with an overly dramatic tremolo. Nowadays I disagree with my former self on the fact that our relations would be self-evidently founded on exchange. This is because exchange is possible only on the condition of excluding the parasite, which makes exchange derivative of a more basic relation, the parasitic one. The exchange of words we are having now, for instance, is possible only on the condition that we manage to struggle together to exclude noise and all potentially intervening third parties.[11] The parasite is our 'common enemy', a common nuisance we have to get rid of. In this light, then, the collective 'is the expulsion of the stranger, of the enemy, of the parasite'.[12]

And just look at yourself: you are a parasite in your own right. Here I was, in tranquil peace, beginning to tell my audience about the parasite as an untidy guest, but you interrupted me before I even got properly started and immediately kicked up a fuss.

– Now you are being rude, accusing me of being a parasite! I am an acknowledged scholar and have published more than 200 articles in esteemed international...

– Could you please let me go on? What I was going to do, before you interrupted me, from the very start was to weave theorizing together with storytelling and discuss the relations of hospitality, the paradise imagery of tourism and the parasite through the film *The Beach* (2000) directed by Danny Boyle and based on a best-selling novel of the same name by Alex Garland.

– Yep, seen the movie a couple of times. Cannot see what it could possibly have to do with parasitism, though. And making use of fiction in an academic text? Come on...

– Let me assure you that the movie is fascinating in relation to parasitism. While the word 'parasite' comes up only three times in it, one can nevertheless argue that the whole film is marked with parasitic relations. In a sense, the film maps a system of the relations of hosts and guests, giving and taking. It is as if everything in the movie, down to its very production process, as you shall see, was a question of hospitality and the parasite. By engaging with the film, my purpose is to get at the constitution community, and show how it is closely connected to the themes of hospitality, violence and parasitism. To me the movie contains important narratives, impressions and ideas about the relations between host and guest and about the role parasitism plays in those relations. Overall, I think that sometimes fiction may enrich and intensify concepts. However, I am also of the view that we do need philosophy or scholarship to tell us what it is exactly that is so profound in fiction.[13] And, since it is very difficult to watch a movie in a text, I would like to *tell* you what the film is about, instead. Do you mind, my literary opponent?

– OK, then. My eyes are hurting anyway from staring at the computer screen. I promise to hold my breath for a while and listen. I also retain the right to remain suspicious of what you say.

– Fair enough, I will not judge you for that. Let us begin with the movie:

Exiting the bedchamber world

The Beach is a travel story of the pursuit of pure, pristine nature, of finding one's way out from the modern world soiled with litter and noise to a serene paradise. The movie begins with a scene from the streets of noisy Bangkok by night, with the main protagonist Richard, a young American backpacker played by Leonardo DiCaprio, strolling aimlessly from entertainment to entertainment, from water balloon wars to a drinking contest where the contestants are challenged to drink snake blood. Richard has left his ordinary life and old world in order to get to another one. And so he has landed in Thailand. But, as soon as he arrives, he realizes that just crossing the ocean does not suffice. This is because the comforts of his home have followed him on his way, along with the like-minded fellow American homebodies:

> you cross the ocean and cut yourself loose. … The only downer is, everyone's got the same idea. We all travel thousands of miles just to watch TV and check into somewhere with all the comforts of home. I just feel like everyone tries to do something different, but … you always wind up doing the same damn thing.

People wish to go out to see the world without being deprived of the comforts of modern life. And therefore, no one ever actually leaves indoors. Today, we're packed into cities. Our life takes place in homes, offices, banks, bureaus, supermarkets, cars, airports, airplanes, buses, trains, metros, stations, lavatories, nightclubs and hotels. People, as Serres sarcastically puts it in *The Natural Contract*, are '[i]ndifferent to the climate, except during their vacations when they rediscover the world in a clumsy, arcadian way'.[14]

But Richard wants to escape the fall into repetition. He wishes to leave indoors, cut former relations and get in contact with the Real, the real world. In the voiceover narration to the opening scene, Richard says he is 'looking for something more beautiful, something more exciting, and yes, I admit, something more dangerous'. Our ordinary world is forgiving and safe. Even when we rediscover the world in a 'clumsy, arcadian way', it must be cleared up for us in advance. Chaos and disorder must be excluded and order must be created within. In the film, Boyle underscores this nicely in a scene where Richard, having just checked into a hotel, enters the guest facilities. The other residents are watching *Apocalypse Now* from a screen. Lieutant Colonel Bill Kilgore (Robert Duvall), shirtless and wearing a Stetson, comfortingly shouts his comrade(s) in the midst of the flames of the battlefield: 'We'll have this place cleaned up in a jiffy, son. Don't you worry'. To draw on Jennie Germann Molz's chapter in this book, the troops could be said to 'camp in clearing'.

Serres notes that our ordinary world is like a 'bedchamber', where 'everything is forgiving, the bed and the pillow, the armchair and the rug, supple and soft. A thousand causes with nonexistent effects'.[15] *The Beach* ties the spread of the bedchamber world to the tourist industry. Environments around the world are carefully staged and modified to create enjoyable sites for the eyes and cameras of tourists[16] – as well as for their bodies lying next to one another on beaches, for instance.[17]

Travellers like Richard blame the tourists for this. For the traveller, tourists are a global nuisance. They are the most despicable of human kind, if not even the ultimate enemy. The tourist is someone from whom the traveller must set oneself apart; the traveller denies the tourist within oneself.[18] As a sucker for the real and the authentic, the traveller has to deny the tourist in oneself, because the tourists just 'want it all to be safe. Just like America', as Richard complains of his compatriots. Any paradise is just waiting to be invaded by tourists and turned into a holiday reserve or an attraction by tourism operators, a safe and soft indoor space, as it were. By transforming everything real and authentic into 'touristy', tourism annihilates and demolishes all that which the traveller lives off and for in the film. Consuming space, the tourist of *The Beach* is a parasite that threatens to eat away any paradise. What is more, the tourist is, paradoxically, a parasite who is welcomed – welcomed because s/he *gives*, brings in the money. Indeed, it is ultimately money that allows the tourist to play the position of the parasite: 'Pay them in dollars, fuck their daughters, and turn it into Wonderland', as Daffy (Robert Carlyle), a manic drug addict travelling with the fake name Daffy Duck on his passport, announces his anti-touristic motto in the film.

Thus, on the one hand *The Beach* echoes anti-touristic attitudes that were prevalent in the tourism studies especially in the 1980s, with authors such as Daniel Boorstin, Maxine Feifer, recalled by John Urry, and Dean MacCannell bemoaning the loss of authenticity.[19] Yet on the other hand, and more interestingly, the film also shows the problems inherent in the very search for the authentic and the real. Lisa Law, Tim Bunnell and Chin-Ee Ong interpret the film as providing a 'cultural critique' of such a quest.[20] Its point is not that there is no such thing as the 'authentic' in the first place but that every para-dise inevitably ends up being para-sited. The parasite is always there, in-between, in the position of the third, interfering and intercepting.[21] It is always on the channel, plugged into the relation. The very act of finding an unspoilt paradise immediately spoils it, as we will see in the next section.

Paradise enclosure

The paradise that *The Beach* exhibits is not a state of future bliss, but an actual place of perfection, contentment and happiness. The paradise pursued by Richard is a secret, secluded island with a perfect, white beach, crystal clear blue water and more dope than one can ever smoke. Analogous to postcards, tourist brochures, guidebooks and travel catalogues, the visualization of the beach in the film corresponds to 'western voy(ag)eurs' idealized imaginings of a pristine, tropical environment'[22] and as such embodies what Urry has called the 'romantic' tourist gaze.[23] Accordingly, Rodanthi Tzanelli has considered *The Beach* as a form of tourism she calls 'cinematic tourism', where the viewers consume and appreciate magnificent exotic scenery.[24]

But there is also something else in the island that is very familiar, isn't there? Without having to force it, the film's paradise island located in the Gulf of Thailand evokes Utopia, an imaginary island in the Atlantic Ocean described by Thomas More in his 1516 book of the same title. Of places, More notes, Utopia is 'the happiest in the world'. The homophone of utopia, *eutopia*, designates a good place, a place of felicity; in Greek, εὖ means 'good' or 'well' and τόπος stands for 'place'. However, while being related to it, the paradise island in *The Beach* is to be sharply distinguished from both Utopia the island and utopia the notion. In the term 'utopia', the prefix u- is derived from the Greek οὐ('not'). Thus, utopia designates, literally, a no-place, a no-where. Utopias are unreal spaces, spaces without any real place in the world. They 'are emplacements having no real place', as Michel Foucault puts it in the piece 'Different Spaces'.[25] In a political sense, a utopia presents an ideal, perfect community, usually one that is projected to the future. The paradise of *The Beach*, by contrast, could perhaps be described as a *realized utopia*. Like utopia, it reverses the miseries of contemporary society but, unlike utopias, it is a localizable, real place. In the Foucauldian parlance, the paradise imaged in the movie could be characterized as an 'other space' or 'heterotopia': it is a place different from and 'outside all places', but it nevertheless can be localized.[26] The paradise turns the nowhere of utopia into a now and here.

A prerequisite of any paradise is that it must be enclosed. The word 'paradise' arrived to English from the French *paradis*, which is derived from the Latin *paradisus*, Greek *parádeisos* (παράδεισος) and ultimately from the Old Persian word *pairidaêza*. The literal meaning of the word *pairidaêza*, compound of *pairi-* ('around') and *daêza-* ('to make, form (a wall)') is 'enclosure, park'. The idea of walled enclosure was not preserved in later usage, yet the word park – which is of the same family as paradise – is from Old French *parc*, probably ultimately from West Germanic *parruk*, meaning 'enclosed tract of land' and refers to a deliberately enclosed area. In fact, to be precise, the word *parruk* originally referred to the fencing, not to the place that is enclosed with fences. What is more, the Greek word *parádeisos*, which was originally used to refer to an orchard or hunting park in Persia, was used in Septuagint, the Greek Old Testament, to mean 'Garden of Eden'. And the word 'garden' comes from Old English *geard*, 'enclosure'. Paradise thus designates a demarcated and finite space. It is a space protected from the openness of chaos by enclosure.

In *The Beach*, this idea is rendered very evident. First of all, the beach island is in the national park and it is forbidden for tourists to go there. Second, and more concretely, the beach is also hidden from view, as it is sealed in from the sea by cliffs. The cliffs appear as a borderline demarcating the inside from the outside. As a border, they have a dual role: while they protect the inhabitants of the abode with their softness, with their hardness they, as Serres notes of borders, also keep possible invaders out.[27] It is only by being cut off from this world by a border that the beach is able to remain a paradise. It must be kept secret and intact. The first time Richard hears about the beach is from Daffy, who is a former resident of the colony inhabiting it. When they share a joint at the hotel in Bangkok, Daffy reveals Richard the secret of the beach, though at the same time he insists that it must also remain a secret[28]: 'See it's like a ... a lagoon. Ya know, a tidal ... lagoon. See it's sealed in by cliffs. Totally fuckin' secret, totally fuckin' ... forbidden. And nobody can ever, ever, ever, ever go there. Ever'. Otherwise the beach would be immediately invaded by tourist-parasites. Tourists would instantly take possession of the island by soiling it with pollution, both hard (garbage, filth, excrement, exhaust from mopeds and cars, etc.) and soft (signs, images, logos, billboards, advertisements, loudspeakers, etc.).[29] In order to remain hospitable and keep on giving, the paradise must remain pure. And therefore it is crucial to keep the parasites out.

It is only later that Richard, at that point already living on the island himself, realizes what Daffy really meant. When paying a short visit to the outside world in order to stock up on food and equipment, Richard is nauseated by the noise of the world he used to belong to. Garish neon lights, bazaars, bars, criss-crossing cars and mopeds howling, partying and vomiting tourists all over the place, prostitution, techno raves on the beach and so on.

> I'd really been looking forward to air conditioning and some cold beer, but when we got to Ko Pha-Ngan, I just wanted to leave again. In one moment, I understood more clearly than ever why we were so special, why we kept our secret. Because if we didn't, sooner or later, they'd turn it into this. Cancers. Parasites. Eating up the whole fuckin' world.

Therefore, it is important to keep the parasites out. At the same time, however, no inside is ever inviolable. No inside is ever fully sealed. All borders have holes, passages, portals and porosities. Through them, things may enter and leave. Because of the existence of pores and holes, no paradise is ever fully secured from parasites. The total elimination of parasites cannot be attained. All attempts at their permanent and absolute exclusion are doomed to fail. It is only a question of time until the parasites find a way in. That Richard himself, a parasite in his own right, manages to enter the island is a token of that. And before he enters the island, the story of the beach already circulates among backpackers as an urban myth. One evening, when Richard arrives at his hut at a beach resort in Chaweng in heavy rain he notices that he has lost his key. Zeph and Sammy, two American surfer guys from the next hut along from Richard's show him hospitality and invite him over to their porch for a beer and a joint. They pass on the story to Richard, without knowing that he in fact has in his possession a map actually leading to the secluded beach.

The map was left to Richard by Daffy, who was already dead by the time Richard found it enclosed in a letter fastened on his door at the guest house in Bangkok. By leaving Richard the map after verifying that the beach must remain secret, Daffy provides both prohibition and access. He renders the island both sacred and profane. He both encloses the island and lets the parasites in. While insisting that the island must be kept intact, that no one can ever enter it, he not only tells Richard about it but also leaves him the map that leads to the island. Instead of keeping the secret of the beach strictly to himself, he reveals it to a stranger, an odd 'travellin wank' he only had formed a momentary alliance with, a union that was dissolved, went up as smoke, when the joint, the object binding them to one another, was consumed. However, it is not far-fetched to suggest that perhaps for Daffy the beach was already a paradise lost. Perhaps the parasites were in fact already there. When on the island, Daffy did his best to heal the island's immunity system, as it were, from an infection. He pictured himself as a sort of medicine man or a physician of the community: 'See I-I was the one that was trying to find the cure. Procurer of the cure. And I said to them: "You've got to leave. You've got to leave this place". But they wouldnae listen'. The immunity system had already failed. The assumedly rightful inhabitants of the paradise – the men and women with ideals – were parasites just like any other, for they too came from the outside. As Daffy tells Richard: 'Ideals, eh? We were just fuckin' parasites! The big, chunky Charlie!' In other words, the idealistic community, as Law, Bunnell and Ong too suggest, had sown itself the seeds of destruction of the paradise its members so keenly tried to protect.[30] When he enters the island himself and invites other uninvited guests to come after him (in Chaweng, he slips a copy of the map under Zeph and Sammy's door as a way of repaying courtesy), Richard is only a thermal exciter accelerating a process that would have happened irrespective of his deeds.

Interestingly, and sadly, *The Beach* itself does not exempt itself from the parasitic relations that it displays. To a certain degree, allegedly the production process of the film was parasitic in its own right. The Twentieth Century Fox film company faced fierce opposition from environmentalists, pro-democracy groups and local residents in Thailand for making prohibited changes to the protected natural landscape of Maya Beach in Krabi's Phi Phi Island national park for the purposes of shooting the film.[31] The paradise exhibited in the film was, paradoxically, a creation of breaching the peace of a natural paradise. The natural environment was substantially modified for filming to appear more 'tropical' for western spectators.[32] According to Thai environmental activists, the company 'bulldozed the beach, removed native plants and planted some 100 coconut trees because the film script called for a perfect tropical beach, large enough to play football on it'. In a sense, then, they created the beach version of a bedchamber through violence. The effects of the actions were severe:

> During storms that hit the area by the end of the rainy season, the environmental consequences already became evident: The sand dunes dug up and stripped from their natural vegetation collapsed and were washed into the sea. The transportation of equipment and fully-grown coconut trees to the island also damaged coral in Maya Bay.[33]

Furthermore, later *The Beach* was mobilized in the marketing and promotion of tourism by holiday providers, with the beach being advertised as the 'Leonardo Beach'. The site became crowded by masses of tourists taking 'The Beach Tours', wanting to visit the pristine and authentic tropical paradise beach they had seen in the film.[34] It made no difference that the beach itself depicted in the film was fictional (for example, the cliffs surrounding the bay were digitally added). So, the consequences of the movie to the filming location were reminiscent of the events taking place in the film: the paradise was spoiled.

Paradise parasited

Richard sets on the adventure to find the paradise island with a French couple – Françoise (Virginie Ledoyen) and her boyfriend, Étienne (Guillaume Canet) – he had met at the hotel. His generous invitation was partly motivated by the prospect of parasitism. When Étienne opens the door, Richard produces a Freudian slip out of his nervousness: 'Hey! You want to take a hike? Uh.. a-a trip, a journey? With y-your girlfriend and me? I mean the two of you and me. Together'. So, in the blink of an eye, the host himself is turned into a parasite. And indeed, later in the film Richard ends up coming in between the French couple. But he plays the position of the parasite even before that, for he couldn't have reached the destination just by himself. As Richard confesses in a voiceover to the occasion of inviting Françoise and Étienne to accompany him: 'I realised that I had absolutely no idea of how I was gonna get there'. Étienne organizes the whole journey: 'tickets, timetables, the whole damn trip'. And yet, Étienne and Françoise are just as dependent on Richard as he is on them, since without him and the map that he has in his possession they would have never found the beach, nor would they have ever known anything of its existence. So, Richard – who is like a cripple – and Étienne and Françoise – who make one blind person – strike a bargain, as it were, by exchanging legs for eyes and knowledge.[35] Together, they form a functioning unity, a system, as if they were one able body: the blind will carry the cripple, who will be the guide. One is a parasite to the other (and together they also exclude other parasites).

After landing on the island, it almost immediately reveals its wonders to the three travellers. In the same instance, however, its system of parasitic relations also begins to unfold before them. Richard, Étienne and Françoise discover an enormous cannabis field, with a size of a football pitch. However, the incredulous joy and amazement caused by the discovery is cut short. Richard, Étienne and Françoise realize that they were not invited, as Richard spots Thai gunmen on guard. Unwilling to listen to the staccato of assault rifles, the trio interrupts the blood feast before it begins: they narrowly manage to save themselves and flee unnoticed, parasites interrupting and driving each other away one after the other. The greatest parasite is the one who expels all the others.[36]

When the trio finally find their way to the beach, they realize that they are uninvited guest there as well:

> I don't know what we expected. People living in a cage, maybe even a few guys in tents. But nothing like this. It was like we arrived in a lost world. A full-scale community of travellers – not just passing through, but actually living here. I suddenly became aware that we weren't even invited.

Richard, Étienne and Françoise are surely trespassing. They are strangers violating someone else's home. Nevertheless, instead of being simply turned away, the intruders are welcomed as guests/enemies, before determining whether they are legitimate or illegitimate arrivals, guests or parasites. They are accepted, though not unconditionally, for the initial reception is followed by an interrogation. The arrivals are taken to Sal (Tilda Swinton), the leader of the tribe. She wants to make sure that there will not be more of their kind coming. That is the unspoken condition for their truce. When Richard shows the map and tells about the death of Daffy, Sal asks, as if casually: 'Do you think he gave the map to anybody else?' Richard answers 'I-I ... No, I don't think so'. And when all three, Étienne, Françoise and Richard reply 'No' to her question whether they have shown the map to anybody, Sal is relieved: 'Good. We value our secrecy'. And she lights the map in fire.

As soon as the guests stop being in the state of coming, that is, as soon as they come and stay, they are no longer guests. Only a couple of days after their arrival, Richard, Étienne and Françoise settle in and make the beach their home: 'This became our world. And these people, our family. Back home was just one place we didn't think about. I settled in. I found my vocation: the pursuit of pleasure'. At first, the site resembles more a tourist resort than a home. As the voiceover narration provided by Richard tells us, 'There was a range of sporting and leisure activities to suit all tastes', anything from zip lines to video games, swimming, playing football and singalongs by the campfire, you name it. The place is a 'beach resort for people who don't like beach resorts'. However, this is not to say that each would not have to contribute to the well-being of the community somehow. The constitution of community, as Roberto Esposito has argued, is intrinsically tied to the *obligation* to give.[37] The inhabitants have some work-related activities and a rudimentary division of labour among them – with three Swedish guys, Richard provides fish for the community, an Italian guy with the nick name Unhygenix cooks for the community, Sal's partner Bugs is their on-island carpenter and some other people are doing gardening, for instance.

Nevertheless, life on the island is rather leisurely. And for a while, it seemed like nothing could disturb them in their blissful happiness. However, bliss has its price. Unviolated, undisturbed happiness is itself a product of violence. It is only made possible by way of exclusion: 'In the perfect beach resort, nothing is allowed to interrupt the pursuit of pleasure', as Richard puts it. The parasites must be kept away. Richard improves his position in the community by killing one of them, a shark that swam to their lagoon. Afterwards, he gets all the attention to himself by telling all others at the longhouse about his dangerous encounter with the shark. Hence, Richard parasites the parasite and becomes the greater parasite of the two.

However, keeping parasites away is a full-day, non-stop business. It demands constant attention. As Serres notes, one has to be constantly on guard, in shift around the clock 'without sleeping, without turning [one's] back, without leaving for a moment, without eating'[38]. What makes the endeavour all the more difficult is that parasites not only keep flowing in from the outside but they also emerge from within. There are several serpents in the paradise, Richard himself being one of them. He comes between Étienne and Françoise. Keaty (Paterson Joseph) intervenes between Richard and his libido by asking him to forget it, but Richard can't help it, the attraction he has developed for the beautiful Frenchwoman already has a hold on him: 'All in all, this really was paradise. Except for one thing. Desire is desire, wherever you go. The sun will not bleach it, nor the tide wash it away'. One night, Richard and Françoise sneak out from the camp to the beach just the two of them and end up kissing and caressing one another passionately. So, the old order was transformed into a new one: Richard, formerly himself in the position of the third, took Étienne's place and forced him to play the third. Étienne was cast off and a new pair was formed: Françoise and Richard.

For a while they we were untouchable in their happiness. But no relation remains unparasited for long. The parasites keep turning up, no matter what. One day Sal convenes the inhabitants to inform them about a situation. A fungus has contaminated some sacks of rice, and because of this parasite they have to make a journey to Ko Pha-Ngan to stock up on rice. There are no other volunteers but Bugs, but Sal says that Richard will accompany her. And in Ko Pha-Ngan, the intimacy between Richard and Françoise is damaged and soiled, as overnight he gives his body to Sal. However, it is not solely Sal who is a parasite here; Richard too plays the parasite, for Sal is with Bugs, the on-island carpenter. Although Bugs had seen this coming, Richard is a parasite more or less against his will. It is not out of passion but out of necessity that he sleeps with Sal. It is a pact. In Ko Pha-Ngan, when sitting at a bar with Sal, Richard is walked in on by Zeph, Sammy and two German girls accompanying them. Richard realizes that he has been busted. Now it comes clear to Sal that he had lied about not showing the map to anyone. He tries to cover the tracks of his lie and claims that there is no beach after all: the beach was just a story, and the map was fake. But Zeph won't buy it: 'You wouldn't be holding out on us, would ya? Let me guess. It's a fuckin' paradise!' And so Richard sleeps with Sal for her silence. It is his return ticket to the island. And so we have yet another duo bound by a secret.

When they get back to the beach, Richard is giving out souvenirs at the longhouse like it's Christmas. Everybody is happy. Everything seems to be as before. 'So I started just where I left off. It was almost like my trip to Ko Pha-Ngan never happened. Almost'. Richard keeps the house of cards standing with two lies, both of which are about to rumble: to Sal he told that Zeph and Sammy do not have a copy of the map, and to Françoise, who has been suspicious, he tells that nothing had happened between him and Sal back in Ko Pha-Ngan. The paradise really starts to collapse in consequence of a second shark intrusion. Two of the three Swedish guys in the tribe are bitten while fishing in the lagoon. And they are bitten badly: Sten is dead and Christo severely injured. The inhabitants bury Sten and hold a funeral, but things can't get to normal anymore. Sal won't agree to Christo's request that a doctor be called to the island to see him. On the contrary, she suggests that Christo should somehow get himself to the mainland and keep their secret by not telling where the incident happened. But Christo is too scared to go anywhere near the water. And so they are stuck with him, for he isn't getting any better. This is the price of their secret. And Christo's moaning and suffering is starting to really get to the others; it is noise – disturbing and interrupting silence that here is equivalent to order (see Soile Veijola's chapter for more on silent communities). The others see Christo – whose name is surely not devoid of religious connotations – as disturbing noise:

> You see, in a shark attack, or any other major tragedy, I guess, the important thing is to get eaten and die, in which case there's a funeral and somebody makes a speech and everybody says what a good guy you were. Or get better, in which case everyone can forget about it. Get better or die. It's the hanging around in between that really pisses people off.

As a result, Christo is expelled from the community. He is deserted into a tent in the middle of the jungle, where he lies in pain, with only Étienne keeping him company and tending to him. If people want peace, they need to banish the parasite. Of course, such an action is itself immoral. The tragedy of peaceful exchange, communication and dialogue is that it is possible only on the condition of the exclusion of the parasite, which is always an act of violence. But, as in so many other cases, the moral problems our actions may involve tend to be silenced by their effectivity: 'It would be a lot easier to condemn our behaviour if it hadn't been so effective. But out of sight really was out of mind. Once he was gone, we felt a whole lot better'.

The ultimate sign of the ruining of the paradise is the arrival of Zeph and Sammy together with the German girls on a neighbouring island. With binoculars, Sal notices that they have a copy of the map with them. Shutting out the parasites becomes Richard's mission. Sal orders him to keep watch in the bushes all day and night until they come and get the map back, no matter what. Unaware of his mission, the other inhabitants are getting suspicious and start talking about Richard. What is he doing all day? He does no work for the community. 'But he steals our food. I'm certain of it', a woman says. 'Idle, sponging, useless prick!' a man continues. Bugs, who had questioned Richard's usefulness for the community from the very beginning, joins in. All along Bugs the carpenter had regarded Richard as a useless parasite, who can make nothing and produce nothing.

At first Richard thought he would starve to death out there in the jungle. But in fact '[l]ife up on the hill turned up to be a big improvement'. He is freed from the obligation to give and to contribute: there is no fishing duty, no gardening, no complex social relationships. He is playing his own game now. Compared with the life in the village, he finds much more exciting things to keep him occupied. The jungle becomes a massive gamespace[39] for Richard. He sneaks around and spies on the gunmen guarding the cannabis field. He likes to fool around: he imagines that he is playing a video-game, running in the woods, using sticks as guns, rolling on the ground, hiding behind rocks and dodging flying lizards. However, Richard has begun to lose it, just like Daffy had. The beach is too much for him. He cannot keep the abundance of input and sensation in control, but it keeps spilling out, just as it had happened with Daffy. He begins to see things. Daffy had already come to him as a spectre in a dream back in the village, but now Richard is having actual conversations with him. Richard becomes obsessed with Daffy.

> This forest was my territory. Retrieving the map, my mission. And [the gunmen], my defenders. I was the only one with the overview of how it all fitted together. The island. Me. Them. The invaders. All connected. All playing the same game. And at the centre of it all, one man: Daffy.

Daffy, the initial donor, now the ultimate parasite, placed in the centre, stealing all the attention, taking up all space. The parasitic cycle is complete.[40]

Richard the jungle warrior regards Daffy as his mentor. He thinks Daffy has led the way, shown him the truth. So he owes Daffy, and Richard reciprocates by living in faith. He won't betray the beach, he won't let Daffy down. He's on the same side with Daffy. In the jungle, Richard hallucinates that he's walking at the wretched hotel back in Bangkok where he met Daffy, and Daffy pulls him inside his room. The room is in the middle of a battleground. Light filters in through the gun shot holes on the walls, and the wind is blowing. 'Viruses, Richard! Cancers! The big, chunky Charlie's eating up the whole world! Out there!' Daffy hands Richard a pair of binoculars and starts firing from the window with a machine gun. 'Down on the beach! Down on the street! Pay them with dollars and fuck their daughters!' Together, Richard and Daffy will stop the parasites. Together, they will keep away the invaders: 'It starts with four, Richard! Four! But they multiply! They multiply! It's time to stop them! Year zero, kiddo!' Daffy shouts and keeps on firing. 'Year zero!' Richard repeats. Year zero is getting closer. Year zero: 'primal chaos, the state of things about to born, ... a nascent state'[41]. Noise – the beginning and the end of it all.

Next, things evolved as if according to a law of nature. The paradise was about to be ruined, and there was nothing anyone could do about it. Having managed to build a raft and paddle to the island Zeph, Sammy and the two German girls get shot dead in the cannabis field by the guarding gunmen. It is only then that it really hits Richard. This time it is no game, nor does it feel like in the movies. This time it's visceral and real. The screaming, powerless fellow travellers are murdered cold-blooded in front of his very eyes. Richard realizes that it is time to leave the island. He heads back to the village to fetch Françoise and Étienne. They arrange to meet by the boat. However, paradise is not only protected from outside, but it is also sealed from the inside. It is almost as difficult to escape as it is to get in. On their way to the beach, Françoise and Étienne are stopped by the gunmen, and Richard is knocked unconscious and taken to the longhouse.

At the longhouse, the marijuana farmer and his gunmen walk in on the residents in the middle of a techno rave. The farmer speaks up. He tries to convince them that he's not a bad guy, but a producer, a provider, a host, a giver of gifts: 'Do you think I want to hurt you? I'm a farmer, that's all. Understand? I work. I send the money to my family. If too many people come to this island, it's trouble for me! I can't work, I can't send the money and my family don't eat!' He laments that the residents have not kept their side of the deal.

They have become hosts themselves by receiving others, strangers, and that jeopardizes the farmer's giving: 'I said no more people. But more people come. And you ... You give them the map!' The times of hospitality, generosity and abundance are over. The residents are to be banished from the paradise: 'Now, you all go home. Forget this island. Forget about Thailand!'

However, Sal refuses. She makes Richard the scapegoat, blaming that it was all his fault, for it was he who copied the map: 'You let us down, Richard. You brought us trouble'. And what happens to scapegoats? Yes, they are sacrificed. The farmer hands Sal a revolver. The deal is that if she shoots Richard, they can stay. By sacrificing him they would atone for their sins, make things right. René Girard suggests that sacrifice is essentially a mechanism of replacement. By way of sacrifice, the community tries to protect its members from uncontrollable, fluctuating violence by channelling violence to a scapegoat, a relatively harmless sacrificial victim. Sacrifice is thereby violence rendered sacred.[42] Sal pulls the trigger, but the gun clicks. The community instantly disperses and deserts her along with the longhouse. Had the gun actually gone off, the consequence would have nevertheless remained the same: either way, Sal's action forced everyone to see, as Richard says, 'what it takes to keep our little "paradise" a secret'. While a paradise excludes violence it is also based on violence, and Sal pulling the trigger made that explicit. Together, the former residents flee the island. They build a raft, swim back to the mainland and depart ways. The film ends with a scene from an Internet café, where Richard notices that he has received an email from Françoise. The message contains a photo of the beach community, with a handwritten inscription 'Parallel Universe. Love, Françoise x'.

Conclusion

The critical sociologist speaks up, after letting me monopolize all the noise for so long:

– A nice story indeed, but it is only a story. What, if anything, is there for us to learn from it?

I have my answer rehearsed:

– Against much cherished Western political utopias, no community can be absolutely inclusive. A completely open, inclusive community without any exclusion could not survive. It would crumble and collapse in a minute. If the parasites were not kept out, chaos would break loose. The constitution of community relies on drawing a dividing line between the inside and the outside. A border needs to be set up to establish order within and close it off from the disorder of the outside. Order is possible on the condition that chaos is excluded, while chaos, of course, exists only in relation to order. To maintain order, the gesture of exclusion has to be repeated incessantly, again and again. Community thus bears an ambiguous relation to violence: while it shields us from violence, it is also produced by violence. The peacefulness of any community constitutes itself by and in relation to the violent act of excluding the parasites.

– But is it not crucial to suspend violence somehow? Even if you were right that any peaceful co-existence depends and is backed up on violence, in order to survive, collectives must surely also find ways to suppress and limit violence; otherwise they are doomed to extinction, my critic stresses.

I feel that finally we are on the same page, literally:

– You are absolutely right. Ultimately, violence produces nothing but more violence – here we have the eternal return of the same as the infinite cycle of violence. If we are to stop it, we must take responsibility – the responsibility for the *other* – further than our brethren. For are we really acting responsibly if we care only for our friends and the members of our 'tribe', while disregarding and excluding all others? We have to find ways of being responsible for our fellow beings in a world of strangers.[43] We cannot merely look after and care for those who are familiar and the closest to us, that is, the ones who resemble us and are *like* us in certain respects, but we have to be ready to show hospitality to basically *whomever*.[44]

– And you are saying that the parasite is this whomever?

– Yes, everyone can be a parasite. 'Not everyone minus one, but everyone, absolutely speaking'.[45] What I have not yet emphasized enough is that parasiteness is no fixed quality of this or that creature, nor is it a fixed position. The parasite is a circulating epithet, a token passed over and moving back and forth. It is first and foremost the hospitality given that appears as the marker of the parasite. Therefore, parasitism does not necessarily involve any intention to abuse, but one can also be a parasite against one's will. Hospitality necessarily makes the guest into a parasite, irrespective of whether one wants that or not. To be placed in the position of the parasite may also be unpleasant, since parasitism involves dependency; the guest is held hostage by the host, as it were. It is only by repaying the hospitality one has received that one can liberate oneself from the position of the parasite or hostage. And by so doing one makes the previous host into parasite, to whom one becomes a host oneself. Now, you are the host and I am the parasite. Next, I am the host and you are the parasite. We swap places in a process that in principle can go on infinitely. Overall, it is not always clear, who is the host and who is the parasite/guest. The twofold meaning of the French word *hôte*, which means both host and guest, stresses this nicely; the same word is used both for the one who gives and for the one who receives, the one who invites and the one invited.[46] Hosts and parasites constantly swap places. Serres even goes as far as to ask, provocatively: 'Would the best hosts be the best parasites'.[47]

– Are you saying that we have to do away with the negative connotations of the term?

– Exactly. And I tell my interlocutor that the parasite is not only an obstacle for the relation but also its precondition, but still he is not convinced.

– I cannot see in which sense could parasitism and abuse be a precondition of hospitality, for instance!

– It comes down to the issue of taking. By accepting the hospitality offered the parasite gives the empirical possibility of the event of hospitality. There would never occur any generosity, no hospitality would ever be given, if the gesture of showing hospitality was not accepted by the other. The actualization of hospitality relies thus not only on its giving but on its taking as well.

– Sorry, the critical sociologist says, ready to attack. – But to me your idea of being ready to show hospitality to whomever sounds utterly abstract, utopian, illusory and naïve, even. It is precisely something one can expect from a theorist, from someone not bothered with the real world. All burglars and murderers of the world, please feel welcome! What about protection? Do not tell me you do not keep your doors locked at night!

– Now that you mentioned protection, surely you know that an organism is safeguarded from a disease much more effectively through vaccination than through demarcation and keeping at distance. Immunity is achieved precisely by injecting the patient's organism with a tolerable amount of the same disease it is meant to be protected from.[48] I would say that the same goes – *mutatis mutandis* – to the communal body. Because the total exclusion of parasites can never be achieved, it is better to try to make one's peace with them, give them place.[49] By letting the parasites arrive, one welcomes interruption. One interrupts interrupting the parasites from interrupting. Transforming relations from parasitic into symbiotic ones – that may even present a very precondition of ethics.[50]

– Sounds still pretty idealistic to me. And could you please tell me what you understand by ethics?

– By ethics I mean assessing and creating peaceful ways of existing in the world with others. Ethics is about good life or, more precisely, as Emily Höckert, one of my fellow authors of this book puts it in her chapter by drawing on the Bolivian *vivir bien* ideology, about 'living well between ourselves'. Gilles Deleuze emphasized the importance of distinguishing ethics from morality. Whereas morality amounts for him to constraining rules judging the righteousness of people in terms of what they are, do and should be, ethics is a matter of becoming.[51] In the context of hospitality and parasitism, the mode of living together or being-with implied by the ethics that I am suggesting comes down to suspending violence – indeed, to interrupting interruption and forming compositions *with* parasites. Instead of just excluding the parasites, one tries to find ways of co-existing with them. Such symbiosis is thus a space of novelty, of creating new connections and novel compounds. Let me advise you to read also Alexander Grit's chapter on 'hospity' in this book. An example that Serres frequently takes up is cancer. It is not unusual that the total elimination of cancer, especially when the cells have divided and spread far as if by forming a rhizome, is not possible; the parasite does not go away, but persistently remains, at least as long as the host remains alive. In such a situation the critical task is to learn to live with cancer.

At this, my interlocutor looks pensive for a moment. I continue:

– As another example we could think of bacteria. Inevitably, they make us untidy hosts for others. No matter how thoroughly we clean our belongings and ourselves, we never succeed in killing all the bacteria. And that is good, for they are a vital part of the immunity system of our body. If our skin for some reason suddenly rejected some of the bacteria living on it, we would die. And bacteria can also be used in making wine and cheese.[52] Speaking of wine, would you fancy a glass?

– Yes please. Judged by the way the critical sociologist looks, he is thinking: finally you asked!

– Make yourself at home. Be as untidy as you like. Stay if you like. I might tell you about another film now that we have an understanding of our relationship. I once saw Lars von Trier's *Dogville* in the movies with a friend. I reckon you might like it too – by now. In it, a stranger, a beautiful young fugitive called Grace played by Nicole Kidman arrives at the town of Dogville by accident.[53] Somewhat grudgingly, the inhabitants show her their hospitality, but soon she becomes the host and they the parasites...

Notes

1. The chapter is partly based on previously published material, excerpted by permission of the Publishers from the chapter 'The Parasites' Paradise (Lice Hopping on the Beach)', Olli Pyyhtinen, *The Gift and its Paradoxes*, Farnham, Ashgate, 2014. Copyright © 2014.
2. Michel Serres, *The Parasite*, translated by L. R. Sehr, Minneapolis, University of Minnesota Press, 2007, p. 6.
3. Ibid., p. 6.
4. Ibid., p. 193.
5. Ibid., p. 24.
6. http://www.theonion.com/video/meet-the-man-inside-the-nicolas-cage-costume,27318/, site accessed August 2013.
7. Serres, *The Parasite*, p. 144.
8. Ibid., p. 8.
9. Ibid., p. 79.
10. Olli Pyyhtinen, *Simmel and 'the Social'*, Basingstoke and New York, Palgrave Macmillan, 2010.
11. Michel Serres, *The Natural Contract*, translated by E. MacArthur and W. Paulson, Ann Arbor, The University of Michigan Press, 1995, p. 9.
12. Serres, *The Parasite*, p. 56.
13. See also Pyyhtinen, *The Gift and its Paradoxes*, pp. 9–10.
14. Serres, *The Natural Contract*, p. 28.
15. Ibid., p. 111.
16. Dean MacCannell, *The Tourist: A New Theory of the Leisure Class*, Berkeley, University of California Press, 1976; John Urry, *The Tourist Gaze*, London, Sage, 1990; Lisa Law, Tim Bunnell and Chin-Ee Ong, 'The Beach, Gaze and Film Tourism', *Tourist Studies*, 7.2, 2007, p. 144.

17. Soile Veijola and Eeva Jokinen, 'The Body in Tourism', *Theory, Culture & Society*, 11.3, 1994, pp. 125–151.

18. See also Dann, according to whom the traveller wants to expel the tourist within oneself by projecting tourist-ness onto others and distancing oneself from them. Graham Dann, 'Writing Out the Tourist in Space and Time', *Annals of Tourism Research*, 26.1, 1999, pp. 159–187.

19. Daniel Boorstin, *The Image: A Guide to Pseudo-Events in America*, New York, Harper, 1964; Maxine Feifer, *Going Places*, London, Macmillan, 1985; MacCannell, *The Tourist*. For a critique of anti-touristic attitudes, see Veijola and Jokinen, 'The Body in Tourism'.

20. Law et al., 'The Beach, Gaze and Film Tourism', p. 142.

21. Serres, *The Parasite*, p. 63.

22. Law et al., 'The Beach, Gaze and Film Tourism', p. 155.

23. Urry, *The Tourist Gaze*.

24. Rodanthi Tzanelli, 'Reel Western Fantasies: Portrait of a Tourist Imagination in *The Beach* (2000)', *Mobilities*, 1.1, 2006, pp. 121–142. For more on cinematic tourism, see Tzanelli, *The Cinematic Tourist: Explorations in Globalization, Culture and Resistance*, London and New York, Routledge, 2007.

25. Michel Foucault, 'Different Spaces', in James Faubion (ed.), *Aesthetics, Method, and Epistemology*, London, Penguin Books, 2000, p. 178.

26. Ibid.

27. See Michel Serres, *Malfeasance. Appropriation Through Pollution?* translated by A.-M. Feenberg-Dibon, Stanford, CA, Stanford University Press, 2011, p. 43.

28. I have discussed the relation of paradise to secrecy in my *The Gift and its Paradoxes*, p. 69.

29. See Serres, *Malfeasance*, on the distinction of hard and soft pollution.

30. Law et al., 'The Beach, Gaze and Film Tourism', p. 145.

31. For more on the campaigns launched against the film, see Tim Forsyth, 'What Happened on "The Beach"? Social Movements and Governance of Tourism in Thailand', *International Journal of Sustainable Development*, 5.3, 2002, pp. 325–336; Tzanelli, 'Reel Western Fantasies'; Law et al., 'The Beach, Gaze and Film Tourism'.

32. Law et al., 'The Beach, Gaze and Film Tourism'.

33. http://www.twnside.org.sg/title/beach-cn.htm, website accessed in August 2013.

34. Tzanelli, 'Reel Western Fantasies'; Law et al., 'The Beach, Gaze and Film Tourism'.

35. The assemblage of the cripple and blind person is an adaptation from Serres, *The Parasite*, p. 37.

36. Serres, *Genesis*, translated by G. James and J. Nielson, Michigan, University of Michigan Press, 1995, p. 131.

37. Roberto Esposito, *Communitas: The Origin and Destiny of Community*, translated by T. Campbell, Stanford, CA, Stanford University Press, 2010.

38. Serres, *The Parasite*, p. 12.

39. On gamespace and films, see Steven Shaviro, *Post-Cinematic Affect*, Winchester and Washington, Zero Books, 2010.

40. For more on the parasitic circle, see my *The Gift and its Paradoxes*, esp. pp. 147–148.

41. Serres, *The Natural Contract*, p. 51.

42. René Girard, *Violence and the Sacred*, translated by P. Gregory, Baltimore and London, The John Hopkins University Press, 1979.

43. Kwame Anthony Appiah, *Cosmopolitanism: Ethics in a World of Strangers*, London and New York, W.W. Norton & Company, 2006.

44. Jacques Derrida, *Of Hospitality: Anne Dufourmantelle Invites Jacques Derrida to Respond*, translated by R. Bowlbry, Stanford, CA, Stanford University Press, 2000.

45. Serres, *The Parasite*, p. 117.

46. Ibid., 15.

47. Ibid., p. 111.

48. Esposito in Roberto Esposito and Anna Paparcone, 'Interview', *Diacritics*, 36.3, 2006, p. 51.

49. Serres, *The Parasite*, p. 88.

50. For Serres's ethics, see also Julian Yates, '"The Gift is a Given". On the Errant Ethic of Michel Serres', in Niran Abbas (ed.), *Mapping Michel Serres*, Michigan, The University of Michigan Press, 2005, pp. 190–209.

51. Gilles Deleuze, *Negotiations. 1972–1990*, translated by M. Joughin, New York, Columbia University Press, 1995; see also Daniel W. Smith, 'Deleuze and the Question of Desire: Toward an Immanent Theory of Ethics', *Parrhesia*, 2, 2007, pp. 21–36.

52. I've discussed the examples of cancer and bacteria also with Turo-Kimmo Lehtonen in Pyyhtinen and Lehtonen, 'Michel Serres ja yhteisön logiikat [Michel Serres and the Logics of Community]', in Ilkka Kauppinen and Miikka Pyykkönen (eds), *1900-luvun ranskalainen yhteiskuntateoria*, Gaudeamus, Helsinki, 2014 (accepted for publication).

53. For an analysis of the movie, see Gerald More, *The Politics of the Gift: Exchanges in Poststructuralism*, Edinburgh, Edinburgh University Press, 2011.

4
Towards Silent Communities

Soile Veijola

Introduction

We wander around in a semi-urban neighbourhood as if we were a herd of reindeer. We have agreed to follow the set of instructions given to us: to keep silent, keep moving, turn off our mobile phones (and not peek at them even to check the time) and, most importantly, stay together. We are not supposed to choose a leader or a destination. In our backpacks we carry some 'reindeer food'; we can stop to eat if we like. Our instructors accompany us at a short distance, playing the part of 'reindeer dogs', in neon-yellow vests, keeping track of the time and looking out for our safety should we happen to cross a road or encounter curious people who ask us what we are doing. The instructors would explain to them that we are, yes, pretending to be reindeer. One of them carries a reindeer bell, which tinkles as he walks, telling any of us at the head of the herd how far back the last reindeer is.

Step by step, hour by hour, the autumn afternoon darkens into dusk, and the assemblage of streets, blocks of flats, gardens and parks – organized in terms of human categories of city planning – becomes a plain material environment with varying surfaces, curves and corners that allow, direct and obstruct our *Wanderung*. I move along as part of the herd and, when a break – which any of us can call – comes along, I lean against a tree in a park or lie down on my back on the grass in the yard of a random blockhouse, watching the clouds passing by between the branches. I withhold my impulses to try to steer the herd in specific directions (I know there is a forest

68

nearby for instance; as it happens, the event is taking place in my very neighbourhood). Likewise, I try not to analyse my experience quite yet, focusing instead on mere awareness and presence, on simply being part of the herd. I follow the movement of the other reindeer with my imaginary 'reindeer eyes', which, you perhaps do not know, are able to see 180 degrees forward and backward on the left and right but leave a blind spot right in front.

What appeals to me most in our slow moving around as a pack of strangers – almost like a group of tourists, but not quite – is that what we are doing we are doing *without speaking*. It allows me to focus on my own experience while being with others: I am alone and together, free and dependent. I do not ignore the others even though I do not seek eye contact or a chance to talk with anyone. I am oriented both *toward* and *withward* all the others in the herd.

At the end of the experiment, our guardians kindly gesture us onto a small strip of grass at a crossroads, asking us to walk in a small circle; only after a while are we asked to stop – and welcomed back to the speaking community. We feel sad and reluctant to return, and start talking and interrupting each other in our eagerness to share the loss we have just become aware of.

***

The event was an experimental, participatory performance called *The Reindeer Safari*, created by a Helsinki-based live-arts collective called *Other Spaces*, which visited Rovaniemi, Finnish Lapland, in September 2013. The group invents and develops collective, physical exercises by which people can try to experience the modes of being of other beings[1] than humans. In the course of its eight-year existence, the group's exercises, initiated by theatre theorist Esa Kirkkopelto, have become increasingly political and ecological. This particular performance – whose name, *The Reindeer Safari*, invites misunderstandings in Lapland, where the same term is used for a very different activity marketed to tourists – seemed to question the ways in which we perceive being, togetherness, time and place.

In what follows, I try to understand why being a reindeer (as if there is such a thing as *a* reindeer) was such an inspiriting experience for me. Being a bit of a surrealist, I like its association with the surrealist *errance*, aimless wandering, during the 1920s as well as the situationists' *dérives* of the 1950s and 1960s; these experiences were identified by Emma Cocker in her analysis of the relations between life and art and between speed and stillness in performance art in urban settings.[2] My aim is somewhat different: to understand better the relations between life and *the social* by studying the interrelations between *silence* and *community*. For the task, I will 'arealise' my scholarship, to borrow the term from philosopher Jean Luc Nancy, and deliberately trespass fixed theoretical boundaries if I see an inviting opening somewhere (what recklessness one learns in a reindeer herd!) in order to explore the *ontology of being as ethical togetherness*.

'In the company of Nancy, I am able to bring my broader question – of the ways in which "one" and "many" are constituted in our discourses on *being-with* in philosophy and sociology – into the field of tourism research'. Specifically, my ambition is to understand the social relations between and among tourists and the locals as ethical potentialities rather than as merely economic opportunities for both parties. In tourism theory, the question boils down to the host–guest relation, which has been transformed since the early days of tourism and its analyses in sociology and anthropology[3] into a complex system of mediated, essentially instrumental and industrial relations. Tourism today leans on the notion of a modern, mobile, individual subject being served and charged for services, en masse, by the globally institutionalized tourism and hospitality industries. For Nancy, however, the modern individual subject is positioned differently: it is a remnant, an experience and an abstract outcome of the process by which the community is being destroyed.[4]

In Nancy's view, being-with (*être-avec*) and most importantly being-with in plural (*être-avec-à-plusieurs*) is the ontological and ethical foundation of being. The notion of 'avec', 'withness', is thus *not* something to be *added* to being; it *equals* being. In his thought, being, as well as singularity, is plural by definition: there cannot be a singular being without other singular beings. A single being is thus a contradiction in terms; a single being cannot 'be its own foundation, origin and intimacy'; 'one equals more than one, because "one" cannot be counted without counting more than one'.[5] Ontology for Nancy is thus 'a co-ontology',[6] and *being singular plural*, as he phrases it, therefore means that the ontological, the social and the ethical have the same origin. Or, to play with prepositional relations, '[t]he one/the other is neither "by", nor "for", nor "in", nor "despite", but rather "with"'.[7]

The sociologist Georg Simmel[8] also took being-more-than-one as his starting point and, as Olli Pyyhtinen[9] points out, did so even before Nancy and Martin Heidegger[10]. In Simmel's 'sociology of association', 'our existence is essentially coexistence'.[11] For Simmel, '[b]eing-with is not added to nor does it come second to being, but being is already given as being-with; being-with constitutes the being of individuals'.[12] Simmel was not too keen on studying the *embodied* nature of being, though, whereas Nancy views the singular subject as an embodied, finite being exposed to other singular beings in mutual presence and proximity. The embodiment of being is the key to the experience of being in this world; it is always open and exposed to the unexpected and thereby unavoidably vulnerable.[13] In addition to being vulnerable, bodies can of course, to follow Baruch Spinoza and Gilles Deleuze, come together 'in agreement'; they can dwell in 'joyful encounters' with other bodies and have 'common notions' of the world.[14]

Let me now rephrase my question with the previously defined starting points in mind: How could *plurality as an embodied being-with* be addressed in tourism theory in ethically, ontologically and experimentally inspiring and consequential ways? In less philosophical terms, one might ask: What is the role of *the others* (not *The Other*, this time) in tourism and tourism research?

I find these questions to be far from trivial and equally far from abstract. The world is extensively exposed to tourism and tailored to meet tourists' needs. Tourists are everywhere – but not necessarily 'together'. Moreover, contemporary mobile life has an overarching touristic dimension. The social, cultural and ecological impacts of the growth of tourism, with more and more of us searching for 'another way of being and doing' by visiting places away from home more and more often, makes the question based on Nancy's ontology – how to be and live together ethically – even more urgent. Whilst in Nancy's framework being is founded on being-with and the ontology of being is ethical from the very outset, the mainstream research tradition of tourism and hospitality studies has regrettably not addressed plurality in a similar fashion. Rather, plurality is epitomized through a separation between self and quantified others – be they locals, tourists, tourism entrepreneurs or their employees. The convention has left hardly any space for thinking of 'withness', the 'in-between', since it has proceeded from subjects as objects of sales and marketing – in other words, as 'the masses', without relations in-between.

The incessantly reproduced cultural juxtaposition between 'individual travellers' and 'mass tourists' is captured astutely in a painting by Caspar David Friedrich from 1818 depicting an aristocratic man looking down at the sublime view of a mountain landscape in the sacred and possessive singularity of his solitary, romantic tourist experience. The image stands in sharp contrast to the illustrations of mass tourism in modern society: staged golden sand beaches looked upon with an instrumental eye by hotel owners, regional authorities and airline industries expecting profit through volume in high seasons. In between one and many, and in between one and the environment, *the social* – which I would like to define here as embracing the institutionalized, spontaneous, embodied, material and symbolic *forms of being-with-strangers* – has not been a central subject of study in tourism and hospitality theory.[15]

However, if we understand plurality as 'the foundation of being', what kind of a tourist experience – and host experience – would call for and allow being true and ethical to both oneself and to others? With tourism being an incessantly growing form of contemporary mobilities and hospitalities globally, the question of plurality is given an extra twist: How can ethical relations between *strangers* be thought of and, most importantly, practised *without* imposing the idea of *a community* on either the hosts or the guests? The concept of community might bring to mind, on the one hand, ancient villages, patriotism and local interests – facets listed by Roberto Esposito[16] – but also, on the other hand, the technology-mediated 'virtual communities' of today – explored by Jennie Germann Molz[17] – that spring up among people who do not need to have personal or place-bound strings attached to their connectivities. Neither form of community is, however, timed and placed quite in the same way as 'tourist communalities', which typically blossom in holiday resorts or in hobbyist gatherings, lasting for only a week or two; they resemble a *communitas*, a breaking of the subject boundaries described by anthropologist Victor Turner.[18] In the tourist forms of a *communitas*, it is perhaps not so much the *ethics* that stands out as characteristic of tourist encounters but the carnivalist aspect of the temporal suspension of the structures of everyday life.[19]

An alternative reading of 'the tourist community' is that presented by Bülent Diken and Carsten Bagge Laustsen[20], followed by Claudio Minca;[21] all of them compare it to *a camp* in the biopolitical framework of philosopher Giorgio Agamben. For Agamben, a camp is 'a state of exception beginning to become the rule'; it is 'a temporary suspension of the rule of law on the basis of factual state of danger [that] is now given a permanent spatial arrangement, which as such nevertheless remains outside the normal order'.[22] (See also the Introduction and the chapter by Jennie Germann Molz in this book). The approach of Diken, Laustsen and Minca, however, overlooks the sociological analysis of actual being-together in the tourist resort, that is, the forms of the social and sociabilities in it. Indeed, walls and bracelets, as well as breaking the norms of everyday life bear a (faint) resemblance to a refugee camp, but the theoretical richness of the notion of 'a tourist community' (as well as the agony and ontology of a 'refugee community') is effectively lost when looking down on tourists from above rather than being-with-the-other-tourist oneself.

A more general question, however, is whether the concepts of an *autonomous agency* (an undivided individual), on the one hand, and *a community based on belonging and identity* (or lack thereof),[23] on the other, are of much use when thinking of ways of living together and visiting each other in the world of the future. Will they last as the foundation of social life and its theorizing? Is there an alternative way to 'form a community without affirming an identity' and to show 'that humans co-belong without any representable condition of belonging',[24] as Agamben puts it? My intuitive, and undoubtedly only partial, suggestion for an alternative being-with is *a silent community*. As opposed to experiences of both the carnivalist *communitas* and the privatized, individualist modern lives that are stored, cyclically, in all-inclusive holiday resorts, the notion of *a silent community* can hopefully help us re-envision one's possibilities of being-with-others *and* being-with-oneself without an ethical split between tourism and everyday life.[25]

In what follows, I will restage my wandering-as-a-reindeer in the reality of words, in a book we are writing and reading both alone and together, in silence. Simmel, Heidegger and Nancy have alerted me to the ways in which togetherness, being and the world are linked. But to gain an understanding of being-silent-with-another and the seemingly paradoxical (and potentially violent) notion of 'a silent community', I need help from other thinkers as well. I anchor my chapter to our book by asking: How can tourism and its research be *sustained*, ontologically and ethically speaking? More specifically, is it possible for embodied human beings to be 'in agreement' in presence and in the present tense without speaking? What are 'the conditions of possibility' of silent communities and on what terms do they neighbour their outside worlds?[26]

Disrupting silence

Silence as an experience – quite popular among urban dwellers when dreaming of a holiday – is often understood as the opposite of being *disturbed by noise*. By noise I refer to sounds interrupting or putting an end to one's focused or idle being and doing. Semiotically, if we think of why silence is meaningful for people, this is not quite true, however. The opposite of noise would be (not silence but) total *soundlessness*, or, in practice, death or planetary solitude, such as that depicted tellingly in the closing scene of the movie *Space Cowboys*, where Tommy Lee Jones, the cosmonaut, is staring at the stars. What semiotics shows us is that the social meaning of silence cannot be grasped using a scale (here, measuring the sonic space) on which soundlessness gradually morphs into noise or noise into soundlessness. Instead, the cultural meaning of silence as a soundscape to be experienced springs from the *simultaneous prevailing* of the *negations of both*: when non-soundlessness and non-noise co-exist, they create an acoustic environment that is experienced neither as disturbing noise nor as isolating death.[27] We hear the sounds of human or natural life but are not in any way bothered by them. Indeed, were we to follow philosopher Michel Serres, we might say that 'background noise may well be the ground of our being ... it is limitless, continuous, unending, unchanging. It has itself no background'.[28] Thus, life *is* and life *makes* noise. While life makes noise, noise makes life.[29]

As said, rather than being seen as auditive soundlessness, the human experience of silence has mainly been understood as a matter of subjective experience, equivalent to 'stillness'.[30] Hearing non-disturbing and non-alarming sounds ('the noise of life') contributes to *an experience of security of existence*: one feels consistent and imperishable in one's being since one is neither *isolated from* 'the noise of life' or *threatened* by any frightening noise made by others. Importantly, even in being with others, one is able to act upon the world freely.

One way to understand silence is, indeed, in terms of an intersubjective state of being-with that furthers an experience of both safety (continuation of existence) and freedom (the possible, the potential and the serendipitous). The former idea is captured in the famous image created by D. W. Winnicot[31] of a baby sitting on the floor engaged in play while her mother sits in an armchair nearby. The baby experiences a mode of being-with that does not interrupt her being with herself: the baby can immerse herself/himself fully in her/his imaginative world of play. The experience may be mutual; a child lost in play allows the mother – if she is not anxiously attached to her baby – to get lost in her own thoughts for a while. Neither interferes in the time and space of the other.[32] Winnicot calls an experience of this kind an experience of *potential space*. It could also be seen as a state of *being alongside* one another – as Joanna Latimer puts it, 'attaching and detaching to different others, partially connecting and partially disconnecting' and thus producing 'a form of dwelling amidst different kinds'.[33] Both experiencing potential space and being-alongside can be understood as ways of being-with, with a twist toward the social.

Andrew Metcalfe and Ann Game understand Winnicot's concept of potential space as an open way of being, a non-finite difference emerging from a relation.[34] It is not the subjects who create or are created by the social, nor is this relational condition to be understood as an emotion felt by a subject. The notion of potential space urges an ontological shift from the logic of outside/inside to the relational logic of potential space.[35] During the reindeer walk, the environment certainly was part of our being-with. It was almost as if we were babies setting out to investigate our environment, although equipped with the able bodies of adults. (Well, in this respect, we were no different from the tourists by the pool sides of the all-inclusive hotels.) Moreover, it was blissful not to need to hear others chatting or having a (one-sided) telephone conversation with people who were not there. It also felt relaxing not to have to show, explain or comment on anything to help the others get the most of their tour or to be responsible for their moods, to *hostess*.[36]

In semiotic terms, another aspect of the social meaning of silence could be articulated with the help of a deontic square,[37] which lays out the structure of meanings in the *modal* dimension of *obligation* and *having-to*. In Winnicot's scene, the child is not *obligated* to play, *nor is it forbidden to*. Instead, the child is both released (non-obligation) and permitted (non-prohibition) to play[38] – that is, of course, if the mother allows the baby to engage in her/his own play without the mother's well-meaning or nervous interventions, and if the baby is not afraid that the mother will disappear at any minute. In a state of silence, the baby is able to be-with-oneself and be-with-the-other, simultaneously.

Here, we can see silence entailing the relational logic of potential space, which disintegrates the juxtaposition between being-one and being-many. The other *can* be a part of the experience of silence. The way in which the child and the mother 'mind their own minds' while being together emphasizes how unfruitful and violent the semiotic opposition of, say, *responsibility* and *freedom*, ultimately is – at least for beings like us who inhabit our world with others. Again, total responsibility and total freedom as modes of being may not prevail *simultaneously*, semiotically speaking, but their negations may, in which case we are *both non-responsible and non-free* vis-à-vis one another.

The latter is perhaps what Mihaly Csikszentmihalyi[39] meant by his concept of 'the optimal experience' of *flow* which could be described as the experience of being completely absorbed in one's own being or doing; being part of a whole in which one's goals make sense and being in the centre of pursuing them; this centre is continuous with the environment of one's being or doing. The experience of flow can occur in any kinds of activities and performances, from sliding down the slopes on a snowboard to painting a landscape or solving a sociological problem – but only when engaging in these activities for their own sake, not in pursuit of any external goals. Interestingly, experiences of flow and of silence can occur in both stillness and movement, both on one's own and with others. Flow is an experience of forgetting one's self and silencing one's inner noise, mental and emotional, fearful or denigrating, when merging with one's doing. When in flow with others, 'bodies' are indeed 'in agreement', having 'common notions'.[40] 'To flow is to be as happy as a human can be', as Turner succinctly puts it.[41] In flow, we do not bear the weight of the world on our shoulders; we are part of that weight, which becomes light, if only for a moment.

The dilemma for many of us – adult beings living in contemporary westernized society, marked by relentless capitalism – is that we are rarely able to experience a similar kind of mixture of freedom and belonging in our relations and co-presence with other people and environments, excluding (in this respect, happy) families, teams and friendships. Working life, especially, tends to be marked by the exact opposite: one is constantly constrained, interrupted and alerted by an incessantly changing agenda of urgent issues set by others from within their time, place and perspectives. We receive text messages, phone calls, emails, Facebook posts and walk-in visitors. Unless we hide away beyond the reach of information technology entirely and flee from our offices, we find ourselves in a *permanent state of interruption* – on hold, waiting, alerted, disturbed.[42] *Interruption* has become 'the permanent state of exception'[43] of our 'bare life'.[44]

Unlike when we are in a state of flow in co-presence with others, in a state of permanent interruption we are not so much subjects-being-together but knots in cords that can be pulled, tightened and dropped by anyone at any moment.[45] We cannot control the timetables of our workscapes since the latter are woven into those of others and managed by way of 'joint custody'. Indeed, there is no going back to quill pens, ink and waiting for the carrier pigeon to bring the reply in a week or two. Yet the flipside of on-going connectivity can be seen in the interruptions by which our minds and time are being possessed, and through which we are being kept on hold, in a state of waiting for the next interruption – while, of course, incessantly interrupting others in our desire to 'stay connected'.

The carousel of contemporary connectivity perhaps explains the increasing popularity of 'withdrawals from the world' to isolated retreats of silence such as monasteries and other camps of silence; these are often seen as the only way to 'interrupt the interrupting', offering silence as an experience that promises a pause in incessant talk and communication. The meaning of a holiday or a sabbatical for many is no longer so much to intensify life by means of 'extreme experiences' – although these desires are of course still there and used to market and motivate travel – but to downshift. Silence, cast as a prohibition against speaking or an agreement *not* to speak, seems to be a venue for an experience of 'potential space', togetherness without imposing. In fact, already in the history of early Christian asceticism, valorized by sociologist Judith Adler, 'silencing worldly thoughts, or "dying to the world"' in the form of a *'peregrinatio* [voluntary expatriation] promised escape from ... corrupting human interchange'.[46]

However, *if* the social is understood as tantamount to speaking and talking, is there a chance for silence as potential space *within* 'the social'? Ironically, at the time of early Christian asceticism laments were heard about the deserts being crowded with monks – and their followers.[47]

It bears adding here that the notion of a monastery as a haven of silence has its linguistic roots in the word 'monk', which comes from the Greek word 'monos' ('alone'). Yet, notes Serres, in monasteries monks live in and as communities.[48] In other words, there is a long tradition of being silent together, but the tradition stands in contrast to modern everyday life – a life marked by economic values that do not follow the cycles of the seasons and nature's rhythms,[49] but the eternal and pressing present tense of the clock-time.[50]

Disrupting the social

What is the matter with 'the social', then? In sociology, the notion has been a subject of excavations for over a hundred years. Simmel in particular has enriched the concept with his theory of its elementary forms: the relations in between two and three beings. For Simmel, the primary sociological unit is two. In his framework, a third is, on the one hand, an outsider in the relation between-two and, on the other, the prerequisite for being-two (by way of the absence or the possible presence of a third). The third is thus a crucial figure for the social and the beginning of 'the proper social', yet it is only in-two when human beings – having eyes that do not leave a blind spot right in front – are able to exchange a mutual gaze and share an uninterrupted intimacy. The third is an intruder, a parasite, an interruption in the sharing of a moment in-two (see the chapter by Olli Pyyhtinen in this book).[51] While twoness embodies the first synthesis and unification – and also the first separation and antithesis – the third engenders a supra-individual social whole.[52] Thus, for Simmel there are only two, three – and many. We could say that while the third always interrupts my being-two, the second (the other) usually interrupts my being-with-myself. The other is both the prerequisite of being for me as a singular being, as Nancy states, and the end of being-with-myself in the present situation. (Unless the two of us release ourselves into a 'potential space' of some kind.)

Being-in-two has also inspired philosophers – from Martin Buber[53] and Emmanuel Levinas[54] to Luce Irigaray[55] and Jacques Derrida;[56] the four have all focused on addressing and welcoming of 'You' in encounters and in discourse (see the chapter by Emily Höckert in this book). In these constellations, plurality is explored in terms of two subjects facing each other in a situation calling for ethical judgements and choices. In their twoness, there is no space for the social brought in by a third. Yet they can never be just the two of them: language functions as 'the third', mediating their relationship. But with language and talking comes noise: misunderstandings, mumbling, requests to talk differently or on other topics[57] (see also the chapter by Pyyhtinen in this book). As Irigaray phrases it,

> My words are an appeal addressed to him. I am saying to the other that his being and my being have, in some sense, encountered each other. I am calling on him to be attentive to what I have perceived, heeded, received from him, and to give value to the truth which is thus announced to him.[58]

Twoness is thus made of reflection, demands and talk. One has no other direction to turn to but toward the 'you' – while turning one's back on all the others.

In Derrida's famous postulations of, on the one hand, an *absolute*, and, on the other, a *political hospitality*, the situation of being-two is depicted in terms of the host welcoming *any* guest standing at her/his door.[59] What happens *in terms of the social* – material, embodied, affective and inevitably also numbered and gendered social relations – once the door is closed behind 'the *arrivant*' is not part of the story.[60] There is only one moment of welcome but no living ever after. (Here Derrida shows an affinity with the Harlequin romances, which also never try to tackle the time after the spouse is carried over the threshold.)

Why do ethical and social relations always need to be demonstrated with the help of being-two? Why should the third be always interrupting? Why is 'the social' reduced to the linguistic and the discursive, to an understanding of being-with in which only 'I' and 'you' have subject positions, while the third person in both singular and plural is an outcast with no power or position in the discourse, as historical linguist Émile Benveniste[61] claims? Why cannot the third, the fourth and the fifth join in a potential space? Is it because the 'I' always dominates in a 'we', as Benveniste[62] claims; according to him there is no such thing as an equal 'we' without the primary role of the one who utters the word 'we', that is, the 'I'. The 'I', who says 'we', hooks up with a 'you' against 'them'; or, alternatively, it makes a pact with 'them' against 'you'.[63] Nancy would, of course, like us to think of 'us' in terms of 'being-many', as always multiple, simultaneously both one and many, without the forever plotting 'I' turning toward or against others.

During our reindeer walk, there was no 'I' or 'we' constructed by a guide, leader or speaker. (Afterwards, of course, there are constructions of 'we-ness' by our telling the story.) Accordingly, there was no exclusion of 'you' or 'them' either. There was only movement in unison in space, a mutual substitution with one walking ahead or last; in the words of Agamben, 'taking-place of every single being is always already common – an empty space offered to the one, irrevocable hospitality'.[64] This is a relief, not a loss. What remains is the awareness of self and others being connected temporarily at the present moment. Indeed, we are 'conjugated in the third person',[65] both within and outside of the herd (not speaking to one another, and being pointed at and talked about by the passers-by); we have escaped the linguistic and social bind of 'I-you' relationships.

In language, 'the linguistic nature of humans',[66] the social manifests itself, for instance, in a number of synonyms for the word 'silence' to be found in a thesaurus[67]: outlining a subject deprived by others of her/his physical or mental ability to speak. A subject can be *hushed, muted, soundless, speechless, tongue-tied, unexpressed, voiceless* or *wordless*. People *deaden, dumbfound, extinguish, gag, muffle, quieten, stifle, subdue* or *suppress* each other. Silence can indeed become a violent modality of being, a prohibition of self-expression. One could also say, metaphorically, that *systems work perfectly when they run silently*. In this case there is no rupture or disruption of, or space for, questioning, the system. Violence within social relations and organizations can, in other words, hide itself with the help of silence; silence can further enforce violence. As the saying goes, 'Evil grows when good men remain silent'.

The previous line of thought is rather different, however, from the notion of 'an exit of this world' through silence, which many of us, as tourists, are understood to need or deserve when going on holiday or wanting to withdraw from the world to either human-made retreats or nature. Silence as the tourist quest builds on a voluntary desire to be still, silent and subdued. Instead of trying to have an effect on the world, one allows the world to have an effect on oneself; one exposes oneself to the world, trustingly. Often nature plays the role of a 'silent mother' who does not intervene but 'is just there' when one wants to find 'peace and quiet, at last'. As an embodied experience of being exposed to mere being, silence could mean the *absence of loneliness*, instead of *being-without the presence* of others.

With Deleuze[68] and his readings of Spinoza, we could describe this modality of being-with as an *affect* (*affectus*), a good encounter rather than a bad encounter, an encounter that increases my 'force of existing' and intensifies its variation (see the chapter by Alexander Grit in this book). I have the power to be affected by the other, and the other has the power to be affected by me. We can intensify each other's being because of our *capacities to be affected*, not because we *intend* to exert 'power' on each other. Affects are, in other words, about social relations and bodies in agreement – or disagreement (see the Conclusion in this book).

There are situations in which silence as a mutual affect embraces us effortlessly: when playing football, for instance; or learning the steps and the mood of dancing an Argentinian tango (providing no other player or dancer is too eager to pick up on the mistakes the others make); or going fishing or hunting. We move in space in a silent communion – for a moment. But as soon as we cease to engage ourselves with the flow of a physical activity and merely *dwell*,[69] we usually feel obliged to talk to one another, to try to share the state of an affect (joy or sadness) by articulating it or, alternatively, think of something to say in the past or future tense. We may also indicate our unwillingness to speak by turning our back on the other or going away. It is as if we always need to establish 'something third' between the self and the other. If we are not kept-apart-and-connected by any other social activity, then we need to turn 'speaking' into one.

Mere being-together without words or action is impossible – unless, perhaps, we have been married for a very long time, and words no longer come easily, or are necessary. Mere being, mere dwelling, (or mere emptiness, discussed by Germann Molz in her chapter) is hardly ever enough for us to form a common ground, a being-with, a silent community. The musician John Cage famously, and deliberately, experimented with the fact in his composition entitled *4'33*.[70]

Why can we not merely expose ourselves to the presence of others? Why is silence-as-not-speaking such a fragile and precarious state of being-with? One notable exception is the Finnish sauna, I dare to suggest. In the culturally codified social time and space of the sauna – where one hears only the background hiss of steam rising from the hot stones and the crackling of the logs burning in the stove, feels the embrace of the heat and humidity in the dim light of a candle or lantern – silence is socially thinkable and possible. Grunts and short remarks like 'Nice heat!' or 'Great sauna you've got here!' do not need to destroy the silence; they are part of it in a way. But even then and there, in the midst of the pure, naked presence of being, the urge to make conversation is usually too pressing. One asks about the next day's programme or weather forecast, for instance. (The other winces and one does not quite know if it is because someone threw too much water on the stones or what.)

Disrupting communities

The notion of 'the tourist community' would seem like no more than a contradiction in terms when juxtaposed with a traditional village community.[71] As a theoretical notion, it starts to make more sense as soon as it is conceived of as a shared experience of not-belonging and not-staying. It questions the notion of ownership and possession as an inherent feature of building and belonging to a community by owning a piece of land, leaning on a symbolic order and producing new life by birth.[72] Yet the members of the tourist community always belong to an *acoustic community*.[73] Just like people can be allocated to spaces within boundaries, hotels and tourist villages can be built in clearly defined areas. But acoustic spaces are not as easily contained within borders. Silence is both precise (the distant sound of a dripping water tap can destroy it) and leaking. So are sounds: they are heard 'as moving, localized and as setting temporal marks', as Steven Feld[74] suggests; they cross boundaries such as walls, doors or stretches of land.

In her analysis of the intersections of private and public sound-scapes, Meri Kytö notes that 'an acoustic order is built by creating, negotiating and following common rules based on values held to be common'.[75] In her view, our connections with others are formed also within the acoustic sphere; yet, 'dwelling in a joint acoustic space does not require that one knows one's neighbours or has shared experiences with them'.[76] In fact, Brandon LaBelle asks us to under-stand the neighbouring noise, or the noise of neighbouring, 'as an ethical encounter that may engender caring for "the other" or the unknown'; it is a lesson in negotiation skills.[77] When acoustic com-munities allow silence as potential space, ethics and aesthetics may coalesce in terms of 'an ethical rather than obedient engagement'.[78]

As we can hear, auditive silence is usually a deliberate (or forced) *obstruction* (see Introduction in this book) rather than a 'natural' way of behaving in the company of familiar or unknown others. In many informal as well as formal situations, one is entitled and encour-aged to produce sounds by speaking – or by starting one's engines or kitchenware – unless the occasion is dedicated to producing an affect through silent participation, as is the case with funerals, wed-dings, acoustic concerts, religious services and keynote sessions in conferences. In these instances, an agreement or a *constitutive rule*[79] of not-speaking is at play in equal measure to a *regulating rule* pro-hibiting it. In my view, the most fruitful way to understand silence as an obstruction would be to take it as *a deliberate deviation of one's automated reactions and practices* rather than as *governing rules* that automatically call for transgression among the 'free, restless spirits'. Breaking a habit can stimulate creativity, as we demonstrate in the Introduction with the help of the film *The Five Obstructions*. When a child is forbidden to play computer and mobile phone games for a while, for instance, s/he might invent a game herself with the artefacts available, with other children present and with the help of her imagination. When an adult is open to an obstruction such as silence, or gives up taken-for-granted amenities as a tourist, s/he can explore the potentialities in her/himself, in being-with others and in being-with the world.

An acoustic community with an agreement on silence is special also in that, whereas one may easily block *visual* images by simply closing one's eyes, one cannot close one's ears from unwanted sounds as easily without using ear plugs and rendering the world entirely soundless in the process. (If one has practised meditation long enough, one is able to *include* all kinds of sounds and even noise within one's meditation, instead of wanting them to stop.)[80] Yet wearing ear plugs will, of course, affect one's manner of experiencing the safety and security of an unknown environment, and modify one's embodied exposure to space and other people. For women who travel on their own, in many cases exposure means a different thing than it does for men.

In the thought-provoking demonstration by Iris Marion Young, for instance, the way we understand the difference between the phrases of 'throwing a ball like a boy' and 'throwing a ball like a girl' reveals the interrelations of subjectivity, space and gender in a patriarchal society.[81] In their play in especially team games, boys often expand and govern their space in relation to others' trajectories in it through their embodied movement and with the help of equipment, whereas girls, when playing team games, often relate to space and others' movement as something that surrounds and often obstructs them from moving freely.[82] The phenomenological idea here is interesting from the point of view of acoustic space: Does gender matter when 'neighbouring relations are negotiated' in spaces devoted to free time and holidays, in play and games for adults and children alike?[83] (See the discussion on *chora* in the chapter by Germann Molz in this book). Do the negotiations take place on an equal footing between, on the one hand, those who want and feel able to conquer a space without hesitating to 'throw their weight around' and, on the other, those who feel restricted by the possibility of this happening? What role do voices and noises generally play in the willingness to conquer space and in the fear of being intimidated by it? Which party, for instance, finds their co-habited lakeside more revitalizing, water-scooter drivers or silence-seekers? Young men, in groups, appear to dominate as the source of sounds of engine-based leisure activities. Then again, when one on one, in many marriages between a woman and man, at least in Finland, the latter often seeks silent moments together in vain.

In short, what kind of silent communities could women and men design (as recreational areas) and maintain (between themselves) together? The sexual difference that Young articulates with her example of movement in physical, material and intentional space evokes the question of the difference between *possessing* and *sharing* a space – be it physical or verbal. Is the other an *object* of or an *obstacle* to my action? Is the other a mere object of my desire to talk endlessly (to whomever)? Or is s/he the ontological and ethical starting point of my being?

It is easy to kill silence. In a gathering of many, if everyone agrees to be silent together, it takes only one person to break the silence and thereby gain more power to decide the circumstances for the others as well.[84] The situation is familiar especially in hotels, where silent nights are promised yet never guaranteed, since hotels in the modern world are not based on being-together in terms of collective rest and sleep.[85] There is no silence unless everyone is willing to be silent.

Then again, experiencing silence together does not need to happen – silently. When people sing or recite something together, (as in uttering 'om' together in yoga), this is an invitation to a meditative state both individually and collectively, often among strangers. On these occasions, vocalization does not work as a third between one and the other but as an act that makes two or more into one, into one voice, into a collective experience of silence, for a moment.[86]

Silence rests on being both autonomous and connected, on sharing both freedom and obstructions. Silence as an experience of both freedom and responsibility epitomizes the good and bad encounters of being-with. In silence, the other is my condition. I realize that *The Reindeer Safari* was a carefully constructed *frame of obstructions* which allowed me to take a leave from my habitual ways of acting and being sociable in social and public spaces, while preventing others from 'interpellating' me back to my habits. It turned my familiar neighbourhood into 'a site of investigation for questioning and dismantling the normative social structure through acts of minor rebellion',[87] as Cocker phrases it.

Paradoxically, by allowing us to give up our agency for a while and just be part of a randomly roaming herd, untidy and uninvited guests who did not care for hosting or being hosted, *The Reindeer Safari* reminded us that 'we have some agency and do not always need to wholly and passively acquiesce'.[88] Or, following Agamben, we experienced 'an inessential commonality, a solidarity that in no way concerns an essence' but 'scatters' us 'in existence' by way of our mere extension.[89] Our walk breached the conventions of a commercialized tourist excursion.

Disrupting hospitality

In order to bring my literary safari closer to its conclusion and to the ontological grounds of contemporary tourism research discussed in our book, I want to question the taken-for-granted notion of *hospitality*, and need Irigaray to help me with the task. This time she draws the line between 'I' and 'You' differently from the way in which she depicted it in the quotation cited earlier. A less possessive way of understanding self and other is offered in the space of hosting and guesting. Namely, Irigaray's psycholinguistic approach to (sexual difference and) the discourse of love questions a well-known 'performative of being-two', that of saying, 'I love you'. We may think this sentence is the most beautiful thing one can say or hear. Yet, in the utterance it is the 'I' who is the subject, while the 'you' is an object on whom the subject projects an emotion. Irigaray suggests that for creating a mediating space in between these positions of interlocution we need the preposition 'to'.[90] Democracy between two is possible, in her view, only if we change the way in which we address the one we love – by not treating her/him as an object. We need to start saying: 'I love *to* you'. In her words,

> [p]erhaps it is possible for me, thanks to the respect that I feel for the other as other, to articulate both attraction and restraint with respect to him. I go out from and return to myself in order to respect his alterity, and this respect for the other becomes respect for myself, my life and my growth.[91]

Analogously, a psycholinguist reading of the relations in spaces of hospitality can bring forth a new way of welcoming the other and being welcomed (see the chapter by Höckert in this book). Instead of saying 'I host you', which turns you into an object of my hosting (in both morally constituted and commercial hospitality) or erases me from my home (in absolute hospitality), I can say: 'I host *to* you'. Instead of saying 'I am your guest', which makes me your possession – or your master – while I stay in your house or hotel, I can say 'I guest *to* you'. This creates a space (of silence?) in between the host and the guest which allows both of them to hold on to their agency and being while allowing the other to hold on to hers/his. *I* am not the source or the substance of your pleasure.[92] Nor are *you* responsible for my inner silence. We are both responsible for our shared soundscape, though. An acoustic space experienced as silence (not as total soundlessness) is not possible without bodies in agreement, without hosting and guesting our being-with *toward* each other.

In the previous horizon of expectations, the *guest* enjoys being permitted or being freed from the obligation to stay or to leave. The *host*, for her/his part, enjoys being permitted or being freed from the obligation to clear the space for the guest of all signs of the former's life (see Introduction). There is equal respect for the self and the other, the host and the guest. There is no need for gating 'the tourist communities' into all-inclusive tourist resorts. There is no pressure on the local people to turn into 'a space of hospitality' for the visitors (see the chapter by Höckert in this book). There is an attunement to collateral, mobile living without ownership and command.

The main reason why I have linked silence to my musings with theorists from such different fields of scholarship is that silence helps to illuminate the ethical ontology of being, regardless of the epistemological presuppositions of a theory. Silence displays a relation between self, other and environment which affords anyone at all a space within space, time within time and relations within relations.

Evidently, the notion of community under exploration in my chapter changes when it is approached from the point of view of silence and when the latter is understood as a potential space that needs to be *cared for by all* in order to exist. The notion of a silent community becomes *camping silence* (see the Introduction and the chapter by Germann Molz in this book), in which one is constantly exposed to and considerate of others, yet the others can be whoever and come and go whenever; they are not one's responsibility or one's permanent neighbours. A silent community becomes a form of *mobile neighbouring*, 'an ethical potential of in-betweenness of human beings, the potential of which could be tapped in architecture and design' in spaces of hospitality, as has been proposed by Soile Veijola and Petra Falin.[93] Camping together in an ethos of mobile neighbouring disrupts the host–guest relation since both hosts and guests have left their homes: the site of their encounter is shifted from the doorsteps or the reception desk of the host to an open field of possibilities, 'a contact zone'[94] characterized by improvisation and interaction. It is a boundaryless, leaking community grounded on respecting oneself and one's dreams of the good life while respecting also one's environment-with-others. It means acknowledging the others who are moving on one's left and right sides, pursuing *their* dreams – near and far, simultaneously.

Conclusion

I have asked whether plurality in tourism can be understood and arranged as a potential space between-many, that is, as an ethical ontology of being-with unknown others. By way of disrupting the interrelations of silent, social, linguistic, communal and hospitable ways of being-with, I have composed the theoretical notion of a silent community as an articulation of being-with in terms of an ethical sociality. To this end, I have been drawing from – and writing a few layers of contemplation on – a real-life subjective experience of an interactive (silent) environmental art perfor-mance which organized its host–guest relations in ways different from the activities on sale in the tourism market. By replacing the starting points of growth of volume for profit or authenticated origin stories of local communities – both of which are used in making tourism legitimate – with a *Wanderung* in the company of Nancy, Simmel, Adler, Irigaray, Young, Winnicot, Csikszentmihalyi and other scholars, I have drifted toward the less frequented shore where ethical being-with and ethical 'being-well'[95] form the onto-logical foundation of social life in tourism.

Perhaps we might add a fourth 'modality of a tourist experience' to the three (rather solipsistic) forms of 'embodied visualities' suggested by Eeva Jokinen and Soile Veijola: climbing a mountain to conquer space, hiding oneself in a dark cleft to endure pain or shame and searching for a symbolic lap to be consoled in.[96] The fourth mode of being would be walking some stretches of the joint journeys with others in a non-possessive and non-indifferent, ethical plural of silence.

Unravelling the imbalance between the duties and rights, respon-sibilities and entitlements, of the host and the guest with the help of the silent community is an initiative to engage in 'experiments of camping together', in which hosting and guesting converge, that is, when they dissolve into the eventfulness of *living* in *any* place. As Adler has phrased it, 'Encoded in even the humblest of travel performances, we begin to discover collectively constructed philosophies – enacted rituals, rather than professed creeds – through which human beings try to comprehend the limits and freedoms of their lives'.[97]

In experimenting with the notion of a silent community as an embodied form of being-with, I am not, by any means, suggesting that silence as such is the ultimate aim of social and ethical being, nor giving advice on who should be in charge of the soundscape of leisure milieus. Rather, I am proposing that understanding the social foundation and meanings of silence helps us imagine and practice the ethical ontology of social being *as dwelling-nearby*.[98] This is a far more responsible goal than simply demanding more space for oneself or supplying tourists with more and more amenity landscapes that are free of other people. Ethical ontologies such as that of mobile neighbouring afford openness to silence, but also to the numerous serendipities,[99] sociabilities[100] and forms of care[101] that frame the moments and milieus of silence in everyday existence.

What else could silence offer for tourism studies? Silence is also a potential space for creativity, therapy and pedagogy, claim Metcalfe and Game.[102] Human relations do not consist of mere relations in-two but are multiplied and rearranged incessantly, especially in tourist spaces. Therefore, it is worthwhile exploring the ways in which potential space for creativity in the form of silence can arrange itself also when dwelling-nearby-with-many – as tourists often do. The human, economic and ecological potentiality of silence for future hospitalities is considerable in a world that is going to be more and more crowded, noisy and neon-coloured.[103] A phenomenon parallel and related to silence – with an increasing cultural value – is *darkness*, which is also disappearing at the pace at which human habitation is expanding across the globe.[104]

If one ceases to view a community in terms of permanent belonging or incessant mobilities and adopts the view of mobile neighbouring instead, the possibilities for freedom and responsibility co-exist – making it possible to be-with-oneself-and-with-others while respecting one's whereabouts. In the words of Agamben,[105] *ease* is 'the empty place where each can move freely'. For him, 'the only ethical experience … is the experience of being (one's own) potentiality, of being (one's own) possibility'.[106] Through silence one can also become the potentiality and possibility for the other.

***

The reindeer meet the reindeer dogs once more, on the following day in the art museum. This time, the performance artists play the role of a group of behaviouralist scientists and present the results of their experiment in diagrams and figures (they announce, among other parameters, the exact duration, length and the average air humidity of the reindeer walk). We, the former reindeer, are amused to listen to information so detached from our own experience of the event. As the last exhibit, we are shown a framed drawing representing the course of our wandering, produced by means of a GPS application carried by one of our guardians. The drawing resembles the head of a reindeer with two horns. So we did leave a trace behind – an unfinished line,[107] an unmarked territory, an un-followed signpost, an unthought-of tourist experience.

Notes

1. See Gilles Deleuze and Felix Guattari, *A Thousand Plateaus: Capitalism and Schizophrenia*, translated by B. Massumi, London, University of Minnesota Press, 1987, pp. 232–238.
2. Emma Cocker, 'Performing Stillness: Community in Waiting', in David Bissell and Gillian Fuller (eds), *Stillness in a Mobile World*, London and New York, Routledge, 2011, p. 90.
3. For example Dean MacCannell, *The Tourist: A New Theory of the Leisure Class*, Berkeley, University of California Press, 1976/1999; Valene L. Smith, *The Hosts and Guests: The Anthropology of Tourism*, 2nd edition, Oxford, Blackwell, 1989.
4. Jean-Luc Nancy, *La communauté désoeuvrée*, 2nd edition, Paris, Bourgois, 1990, p. 16; see also Jussi Backman, 'Olemisen ainutkertaisuudesta ainutkertaisuuden politiikkaan: Parmenides, Heidegger, Nancy', *Tiede & Edistys*, 2, 2013, p. 120.
5. Jean-Luc Nancy, *Being Singular Plural*, translated by R. D. Richardson and A. E. O'Byrne, Stanford, California, Stanford University Press, 1996/2000, pp. 12, 39.
6. Ibid., p. 42.
7. Ibid., p. 34.
8. Georg Simmel, 'The Number of Members as Determining the Sociological Form of the Group', *The American Journal of Sociology*, 8.1, 1902/1903, pp. 1–46, 158–196.

9. Olli Pyyhtinen, 'Being-with: Georg Simmel's Sociology of Association', *Theory, Culture & Society*, 26.108, 2009, pp. 108–128.

10. Martin Heidegger, *Sein und Zeit*, Tübingen, Max Niemeyer, 1972.

11. Pyyhtinen 'Being-with', p. 109.

12. Ibid., p. 110.

13. Jean-Luc Nancy, *Corpus*, translated by Susanna Lindberg, Tampere, Gaudeamus, 1992/1996; see also Eeva Puumala, 'Politiikan tuntu, mieli ja merkitys. Tapahtuva yhteisö ja poliittisen kokemus kehollisissa kohtaamisissa', *Tiede & Edistys*, 2, 2013, pp. 125–138.

14. Gilles Deleuze, *Essays Critical and Clinical*, translated by D. W. Smith and A. Greco, London, Verso, 1998, p. 144; Cocker, 'Performing Stillness: Community in Waiting', p. 93.

15. For exceptions, see Naomi Rosh White and Peter B. White, 'Travel as Interaction: Encountering Place and Others', *Journal of Hospitality and Tourism Management*, 15.3, 2008, pp. 42–48; David Bissell, 'Pointless Mobilities: Rethinking Proximities Through the Loops of Neighbourhood', *Mobilities*, 8.3, 2013, pp. 349–367; Jennie Germann Molz, *Travel Connections: Tourism, Technology and Togetherness in a Mobile World*, New York and London, Routledge, 2012; Soile Veijola and Petra Falin, 'Mobile Neighbouring', *Mobilities*, online first, doi 10.1080/17450101. 2014.936715.

16. Roberto Esposito, *Communitas: The Origin and Destiny of Community*, translated by Timothy Campbell, Stanford, California, Stanford University Press, 1998/2010, p. 16.

17. Germann Molz, *Travel Connections: Tourism, Technology and Togetherness in a Mobile World*.

18. Victor Turner, *The Ritual Process: Structure and Anti-structure*, Ithaca, Cornell University Press, 1969/1977.

19. For example Nelson Graburn, 'Tourism: The Sacred Journey', in V. Smith (ed.), *Hosts and Guests: The Anthropology of Tourism*, Philadelphia, University of Pennsylvania Press, 1989, pp. 21–36; Soile Veijola and Eeva Jokinen 'The Body in Tourism', *Theory, Culture & Society*, 11.3, 1994, pp. 132–135.

20. Bülent Diken and Carsten Bagge Laustsen, *The Culture of Exception: Sociology Facing the Camp*, London and New York, Routledge, 2005, pp. 112–121.

21. Claudio Minca, 'No Country for Old Men', in Claudio Minca and Tim Oakes (eds), *Real Tourism: Practice, Care and Politics in Contemporary Travel Culture*, London, Routledge, 2011, pp. 2–37.

22. Giorgio Agamben, *Homo Sacer: Sovereign Power and Bare Life*, Stanford, Stanford University Press, 1998, p. 169.

23. Soile Veijola, 'Heimat Tourism in the Countryside: Paradoxical Sojourns in Self and Place', in T. Oakes and C. Minca (eds), *Travels in Paradox: Remapping Tourism*, New York, Bowman & Littlefield Publishers, 2006, pp. 77–95.

24. Giorgio Agamben, *The Coming Community*, translated by M. Hardt, Minneapolis, University of Minnesota Press, 1990/2007, p. 86; see also Alphonso Lingis, *The Community of Those Who Have Nothing in Common*, Bloomington and Indianapolis, Indiana University Press, 1994.

25. Jost Krippendorf, *The Holiday Makers: Understanding the Impact of Leisure and Travel*, Oxford, Butterworth-Heinemann, 1987.
26. I thank Suvi Alt for pointing out this aspect of *The Reindeer Safari* to me.
27. On the logic of a semiotic square, see Georg Henrik von Wright, 'Deonttinen logiikka', in Tauno Nyberg (ed.), *Ajatus ja analyysi*, WSOY, Helsinki, 1977, pp. 147–166; Laurence Horn, *A Natural History of Negation*, Chicago and London, The University of Chicago Press, 1989, pp. 12, 263; Soile Veijola, 'Pelaajan ruumis. Sekapeli modaalisena sopimuksena', in Eeva Jokinen, Marja Kaskisaari and Marita Husso (eds), *Ruumis töihin! Käsite ja käytäntö*, Tampere, Vastapaino, 2004, p. 112.
28. Michel Serres, *Genesis*, translated by G. James and J. Nielson, Michigan, The University of Michigan Press, 1995, p. 13.
29. I am grateful to Olli Pyyhtinen for this thought.
30. See for example Bissell and Fuller, *Stillness in a Mobile World*.
31. D. W. Winnicot, *Play and Reality*, London, Routledge, 1991, p. 105.
32. See also Kirsi Määttänen, 'Sense of Self and Narrated Mothers in Women's Autobiographies', in Satu Apo, Aili Nenola and Laura Stark-Arola (eds), *Gender and Folklore: Perspectives on Finnish and Karelian Culture*, Studia Fennica Folkloristica 4, Helsinki, Finnish Literature Society, 1998, pp. 317–331.
33. Joanna Latimer, 'Being Alongside: Rethinking Relations Amongst Different Kinds', *Theory, Culture & Society*, 30.7–8, 2013, p. 81.
34. Andrew Metcalfe and Ann Game, 'Potential Space and Love', *Emotion, Space and Society*, 1, 2008, pp. 18–21.
35. Ibid., p. 18.
36. Soile Veijola and Eeva Jokinen, 'Towards a Hostessing Society? Mobile Arrangements of Gender and Labour', *NORA – Nordic Journal of Feminist and Gender Research*, 16.3, 2008, pp. 166–181.
37. von Wright, 'Deonttinen logiikka'; Veijola, 'Pelaajan ruumis', p. 112.
38. See for example Anu Valtonen and Soile Veijola, 'Sleep in Tourism', *Annals of Tourism Research*, 38.1, 2011, pp. 185–186.
39. Mihail Csikszentmihalyi, *Flow: The Psychology of Optimal Experience*, New York, Harper Perennial, 1975.
40. Cocker, 'Performing Stillness', p. 103.
41. Victor Turner, *From Ritual to Theatre: The Human Seriousness of Play*, New York, PAJ Publications, 1982, p. 58.
42. See for example Mika Pantzar, 'Future Shock – Discourses Changing Temporal Architecture of Daily Life', *Journal of Future Studies*, 14.4, 2010, pp. 1–22.
43. Agamben, *Homo Sacer: Sovereign Power and Bare Life*.
44. Ibid.
45. See for example Jonas Larsen, John Urry and Kay W. Axhausen, 'Networks and Tourism. Mobile Social Life', *Annals of Tourism Research*, 34.1, 2007, pp. 244–262; Andreas Wittel, 'Toward a Network Sociality', *Theory, Culture & Society*, 18.6, 2001, pp. 51–76.
46. Judith Adler, 'The Holy Man as Traveler and Travel Attraction: Early Christian Asceticism and the Moral Problematic of Modernity', in

William H. Swatos, Jr. and Luigi Tomasi (eds), *From Medieval Pilgrimage to Religious Tourism: The Social and Cultural Economics of Piety*, Westport, Connecticut and London, Praeger, 2002, p. 33.

47. Ibid., p. 35.
48. Michel Serres, *Angels, A Modern Myth*, translated by F. Cowper, Paris, Flammarion, 1993/1995, p. 96.
49. See for example Valtonen and Veijola, 'Sleep in Tourism'; Outi Rantala and Anu Valtonen, 'A Rhythmanalysis of Touristic Sleep in Nature', *Annals of Tourism Research*, Annals of Tourism Research, 47, 2014, pp. 18–30.
50. On tourism in monasteries, see for example Kevin O'Gorman and Paul A. Lynch, 'Monastic Hospitality: Explorations', University of Strathclyde Institutional Repository, 2008, http://core.kmi.open.ac.uk/display/9020089.
51. See Simmel, 'The Number of Members as Determining the Sociological Form of the Group' *The American Journal of Sociology*, 8.1., pp. 1–46, 158–96; Soile Veijola, 'Luku, suku ja sosiaalinen: Taipuuko varsinainen sosiaalinen myös naissuvun mukaan?' *Naistutkimus/Kvinnoforskning*, 4, 1997, pp. 15–16; Pyyhtinen, 'Being-with'.
52. Georg Simmel, 'Soziologie', in *Georg Simmel Gesamtaugabe*, Band 11, Frankfurt am Main, Suhrkamp, 1992, pp. 124, 101; Pyyhtinen 'Being-with: Georg Simmel's Sociology of Association', p. 117.
53. Martin Buber, *Ich und Du*, Leipzing, Im Insel,1923.
54. Emmanuel Levinas, *Ethics and Infinity: Conversations with Philippe Nemo*, translated by R. A. Cohen, Pittsburgh, PA, Duquesne University Press, 1985.
55. Luce Irigaray, *Democracy Begins Between Two*, London, The Athlone Press, 1994/2000.
56. Jacques Derrida, *Of Hospitality: Anne Dufourmantelle Invites Jacques Derrida to Respond*, translated by R. Bolwby. Stanford, CA, Stanford University Press, 2000.
57. Michel Serres, 'Platonic dialogue', in J. V. Harari and D. F. Bell (eds), *Hermes: Literature, Science, Philosophy*, Baltimore and London, John Hopkins University Press, 1982, pp. 65–70.
58. Irigaray, *Democracy Begins Between Two*, p. 117.
59. Derrida, *Of Hospitality*; see also Sarah Gibson, 'Accommodating Strangers: British Hospitality and the Asylum Hotel Debate', *Journal for Cultural Research*, 7.4, 2003, pp. 367–386; Jennie Germann Molz and Sarah Gibson (eds), *Mobilizing Hospitality. The Ethics of Social Relations in a Mobile World*, Aldershot, Ashgate, 2007.
60. Aafke Komter and Mirjam van Leer, 'Hospitality as a Gift Relationship: Political Refugees as Guests in the Private Sphere', *Hospitality & Society*, 2.1, 2012, pp. 7–23.
61. Émile Benveniste, *Indo-European Language and Society*, translated by Elizabeth Palmer, London, Faber and Faber Limited, 1973.
62. Ibid., p. 203.
63. For example, see Veijola, 'Luku, suku ja sosiaalinen'.
64. Agamben, *The Coming Community*, p. 24.
65. Veijola, 'Luku, suku ja sosiaalinen'.
66. Agamben, *The Coming Community*, p. 82.

67. *The Wordsworth Thesaurus*, Denmark, Wordsworth Reference, 1993.
68. Gilles Deleuze, 'Deleuze/Spinoza, Cours Vincennes – 24/01/1978', *Les Cours de Gilles Deleuze*, www.webdeleuze.com, 1978/2008, p. 3.
69. See Martin Heidegger, 'Building Dwelling Thinking', in David Farrell Krell (ed.), *Basic Writings: Revised and Expanded Edition*, London, Routledge, 1978/1993, pp. 347–363; Paul Harrison, 'The Space Between Us: Opening Remarks on the Concept of Dwelling', *Environment and Planning D: Society and Space*, 25.4, 2007, pp. 625–647.
70. John Cage, *4'33'*, Edition Peters No. 6777, New York, Kenmar Press Inc., 1960.
71. Soile Veijola, 'Turistien yhteisöt', in Antti Hautamäki, Tommi Lehtonen, Juha Sihvola, Ilkka Tuomi, Heli Vaaranen and Soile Veijola (eds), *Yhteisöllisyyden paluu*, Helsinki, Gaudeamus, 2005, pp. 90–113.
72. Agamben, *Homo Sacer: Sovereign Power and Bare Life*, pp. 175–176.
73. Barry Truax, *Acoustic Communication*, 2nd edition, Wesport, Ablex, 2011.
74. Steven Feld, 'Places Sensed, Senses Placed: Toward a Sensuous Epistemology of Environment', in David Howes (ed.), *The Sensual Culture Reader*, Oxford, Berg, 2005, p. 185.
75. Meri Kytö, *Kotiin kuuluvaa. Yksityisen ja yhteisen kaupunkiäänitilan risteymiä*, Publications of the University of Eastern Finland, Dissertation in Education, Humanities and Theology, No: 45, Joensuu 2013, p. 117.
76. Ibid., p. 24.
77. Brandon Labelle, *Acoustic Territories: Sound Culture and Everyday Life*, London, Continuum, 2010, p. 83.
78. Cocker, 'Performing Stillness', p. 94
79. See John Searle, *Speech Acts. An Essay in the Philosophy of Language*, Cambridge, Cambridge University Press, 1969/1996, pp. 33–42.
80. I thank Hans Christian Hansen for this thought (personal communication).
81. Iris Marion Young, 'Throwing Like a Girl. A Phenomenology of Feminine Body Comportment, Motility and Spatiality', *Human Studies*, 3.2, April, 1980, pp. 137–156.
82. Ibid.; Soile Veijola,'Metaphors of Mixed Team Play', *International Review for the Sociology of Sport*, 29.1, 1994, pp. 32–49.
83. On women's and men's leisure spaces, see for example Betsy Wearing, *Leisure and Feminist Theory*, London, Sage, 1998.
84. I thank Jennie Germann Molz for this thought.
85. See Valtonen and Veijola, 'Sleep in Tourism'.
86. I thank Jennie Germann Molz for this thought.
87. Cocker, 'Performing Stillness', p. 91.
88. Ibid., p. 91.
89. Agamben, *The Coming Community*, pp. 18–19.
90. Irigaray, *Democracy Begins Between Two*, pp. 106–120.
91. Ibid., p. 112.
92. Eeva Jokinen and Soile Veijola, 'The Disoriented Tourist: The Figuration of the Tourist in Contemporary Cultural Critique', in Chris Rojek and John Urry (eds), *Touring Cultures. Transformations of Travel and Theory*, London and New York, Routledge, 1997, pp. 23–51.

93. Veijola and Falin, 'Mobile Neighbouring'.
94. Mary Louise Pratt, *Imperial Eyes: Travel Writing and Transculturation*, London and New York, Routledge, 1992, pp. 6–7.
95. Gaston Bachelard, *The Poetics of Space*, translated by M. Jolas, Boston, Beacon Press, 1958/1994, p. 7.
96. Eeva Jokinen and Soile Veijola, 'Mountains and Landscapes. Towards Embodied Visualities', in Nina Lübbren and David Crouch (eds), *Visual Culture and Tourism*, Berg Publishers, Oxford and New York, 2003, pp. 259–278.
97. Adler, 'The Holy Man as Traveler and Travel Attraction', pp. 46–47.
98. Martin Heidegger, 'Building Dwelling Thinking', pp. 347–363; Veijola and Falin, 'Mobile Neighbouring'.
99. John Taylor, 'Authenticity and Sincerity in Tourism', *Annals of Tourism Research*, 28.1, 2001, pp. 7–26.
100. Georg Simmel, *Grundfragen der Soziologie*, in G. Fitzi and O. Rammstedt (eds), *Georg Simmel Gesamtausgabe*, Vol. 16, Frankfurt am Main, Suhrkamp, 1999, pp. 103–121.
101. For example Eeva Jokinen and Soile Veijola, 'Time to Hostess. Reflections on Borderless Care', in Claudio Minca and Tim Oakes (eds), *Real Tourism: Practice, Care and Politics in Contemporary Travel Culture*, London, Routledge, 2012, pp. 38–53.
102. Metcalfe and Game, 'Potential Space and Love', p. 18.
103. Noora Vikman, *Eletty ääniympäristö. Pohjoisitalialaisen Cembran kylän kuulokulmat muutoksessa*, Tampere, Acta Universitatis Tamperensis 1271, 2007.
104. See for example Tim Edensor, 'Reconnecting with Darkness: Gloomy Landscapes, Lightless Places', *Social & Cultural Geography*, 14.4, 2013, pp. 446–465.
105. Agamben, *The Coming Community*, p. 25.
106. Ibid., p. 44.
107. Jean-Luc Nancy, *Le Plaisir au dessin*, Paris, Galilée, 2009; Martta Heikkilä, 'Monin vedoin: Nancy piirtämisen merkityksestä', *Tiede & Edistys*, 2, 2013, pp. 139–151.

5
Unlearning through Hospitality

Emily Höckert

Introduction

It is an early November morning when we meet at the bus station in Matagalpa, a mountainous town in the highlands of Nicaragua two to three hours north-east of the capital city of Managua. I am happy to see how many of my Nicaraguan colleagues have been able to come. We have all stood at this station before, waiting for the bus that passes through the little farming town of San Ramón. The younger tourism scholars, Idania, Mónica, Danilo and Claudia, tell about their recent visits to the four tourism communities on the coffee-growing hillsides of San Ramón. They are currently writing their final thesis for their university degree, and the research assignment includes assessment and analysis of tourism development in this area. They point out that the fieldwork has been more challenging than they expected, for they have received no more than a lukewarm welcome from the locals, and even otherwise active members of the community have tried to avoid being interviewed.

Their professors, Gabriela and Elba, try to keep up a positive attitude. However, we can all imagine reasons why researchers, students, developers or volunteer workers, with backpacks full of 'developmentalism', are sometimes treated almost as uninvited guests. Community-based tourism – Mónica says with an air of disillusionment – although loaded with great expectations of poverty reduction, empowerment and intercultural understanding, has turned out to be difficult to promote in practice.[1] But we have something else to worry about for the moment: Will *we* recognize the guests whom I have invited to this site and to this narrative?

I am encouraged by the thought that this time I am going to re-visit San Ramón, not by myself as I normally do but with new travel companions who can hopefully offer some fresh perspectives on why rural communities seem to remain little more than mere objects of tourism and development. I feel emboldened also by my fictional story in the making, which will allow me to weave together the empirical data that I have collected during the past years in San Ramón[2] and the theoretical discussions of ethical subjectivities that I have been reading lately. I think it is about time to put it all into a dialogue. My hopes are high: this very day, a long-awaited, utterly eco-friendly excursion will perhaps help not only me but also others to envision and sketch new ways – even utopist ones – of encountering and being together with 'the other'.[3]

Arrival of the first guests

A taxi arrives and stops in front of us. We recognize quite correctly the man who steps out from the car as Jacques Derrida. His writings on hospitality have recently attracted more and more attention among those of us studying tourism, as we saw in the Introduction and the chapter by Soile Veijola in this book.[4] Derrida hurries around the back of the taxi, opens the door and helps his friend out, seemingly delighted at the opportunity to bring along his colleague Emmanuel Levinas. We utter words of welcome, at which they both laugh heartily. I take their laughter to be a burst of relief at having finally arrived at their destination, or possibly they are simply happy to be reunited after many years. Accepting both of my guesses as correct ones, Derrida explains that he and Levinas have a very special relationship to the concept of *welcome* and the act of *welcoming*; namely, he continues, his interpretation of hospitality is based on Levinas's teachings on what 'to welcome' and 'to receive' should mean. Derrida looks at us all and asks whether anyone of us has ever noticed how Levinas's thought as a whole can be approached from the realm of hospitality – subjectivities, ethics, teaching, everything.[5] He continues emphasizing the significance of hospitality by saying that, in fact, 'there is no culture or social bond without a principle of hospitality'.[6]

Derrida observes that we are not as familiar with Levinas's writings as he had hoped. Hence, he clarifies that a valid entry point into theoretical discussions of hospitality is the necessary but impossible conjunction of the *ethics of hospitality* (ethics *as* hospitality) and the *law* or *politics* of hospitality.[7] While the latter are based on obligations and conditions, the former invites us to think about the possibility of absolutely unconditional, open and infinite welcoming. Derrida stresses that it is between these two conceptions of hospitality that responsibilities must in effect be taken. His talk is interrupted by a man who works at the bus station. He asks where we are heading and kindly offers to guide us to the right bus stop. Realizing what time it is, we begin to worry about my third invited guest. I had hoped she would have arrived with Derrida and Levinas.

We start to look around to see where she might be. I notice three tourists climbing into a charter bus. Two of them remind me of the Finnish sociologists Eeva Jokinen and Soile Veijola, my countrywomen. And what is even more surprising, the man accompanying them resembles Dean MacCannell, the founding father of the sociology of tourism. First, I wonder if it really could be them. Then I understand: their fictive journey to Mallorca in 1994[8] is writing itself into my trip. For a moment, I consider running after them and inviting them to join us. Travelling in the middle of the coffee season offers a unique opportunity to participate in the process of producing coffee all the way from picking, drying and toasting the beans to finally sipping the final product with the locals. I realize that I sound like a tourist guide and that they are most likely not interested in being tourists, but only in writing theories about it. Besides, I should focus on finding our third invitee.

Elba pokes my arm to draw my attention to another direction. And yes, we now see a woman walking towards us leisurely, waving to us. She is Gayatri Chakravorty Spivak, best known as one of the key contributors to feminist, postcolonial forms of cultural analysis. She greets us all, even the man working at the bus station. It very much seems that contrary to my conjecture, Spivak, Derrida and Levinas did not contact each other prior to this trip. We welcome her warmly and ask the usual questions: 'How was your trip?' and 'Is this your first time in the country?' Spivak informs us that this is not her first visit here as *her reading* of subaltern speech, the speech ignored in dominant modes of narrative production,[9] was welcomed as a contribution in analysing the Nicaraguan elections in 1990.[10]

Claudia encourages us to start moving toward the right platform, as the next bus to San Ramón will be leaving soon. While everyone buys a bus ticket, my thoughts drift back to Derrida's introduction on hospitality. I wonder whether something like absolutely open hospitality could ever emerge in contemporary tourism settings, especially given that tourism is an economic activity in which social relations become commoditized. Can this kind of hospitality be ethical? Can tourism be ethical if it is conditional? Although I cannot think of any practical examples that could be placed in the category of unconditional hospitality, I can somehow grasp that this view might be used as an endless inspiration for analytical and philosophical discussions of both tourism and development.[11] First and foremost, this idea brings to mind a nostalgic postulation of how, in the past, back in the good old days, people's homes were always open even to strangers. This assumption goes hand in hand with the essentialist statement that warm and open hospitality can still be found in rural areas around the world. However, I have an inkling that Derrida and Levinas might not find it meaningful to divide individuals into selfish urban dwellers, on the one hand, who lock their doors, and kind country folk, on the other, who always have something baking in the oven for any surprise guests that might drop by.

While queuing for the bus, Derrida further explains why he has been interested in analysing whether the ethics of hospitality could furnish a basis for law and politics within a society, nation, State or Nation State.[12] Levinas, reading the tourist guide to the region, politely prompts Derrida to speak about his approach to ethical relations.[13] Derrida goes on, explaining how Levinas's work has focused on the ethics of hospitality in face-to-face encounters, that is, on *being* between self and other. Levinas cordially allows him to be his spokesman on the next point as well. Derrida wants to make it clear that for Levinas ethics is not a branch of philosophy or something that he wishes to construct or define. Instead Levinas approaches the ethics of hospitality as the first philosophy.[14] For him the ethical subject is that very welcome, that very openness to the other. In other words, the subject is not *freedom*, but *receptivity*.[15] I comment on how this kind of thinking sounds somewhat different from the contemporary discourses of sustainable or responsible tourism. In the latter debates, ethics are primarily perceived as a management tool or as a recipe to determine what is right and what is wrong. What is more, there has been limited interest in questioning the Western tradition built on the distinctions between self and other, subject and object, developed and underdeveloped, and so on.[16]

Bus trip to San Ramón

I am the last one to step up into the colourful, old school bus. Spivak and Levinas sit side by side, and Derrida chooses a place behind them, next to a woman who has a rooster in her lap. I am curious whether he is familiar with the controversial trope, used especially by backpackers, of travelling in the 'chicken buses' in Central America.[17] The music is loud and the bus is quite full. Decorative stickers tell us that 'Jesus is my driver'. I get help lifting and squeezing my bag onto the upper shelf. There are no seats left, so I hold onto the hand-rail attached to the ceiling. My Nicaraguan colleagues are spread out, sitting and standing here and there. Danilo, Idania and Claudia have taken out their notes, and it sounds like Gabriela is phoning home to talk to her children. Elba and Mónica are near enough to the visiting philosophers that they can listen to them talking. Luckily, so am I.

Levinas tells Spivak that he has just arrived from a 'vivir bien' (good life, *suma qamaña*) conference in Bolivia – from thought-provoking discussions about different ways of imagining life. The Andean vision of 'good life' embraces an alternative way of understanding reciprocity and the other, an alternative to the Western, growth-oriented, development thinking that separates people from each other and from nature. At stake here is the political activation of indigenous cosmologies in which all being exists in relation to others and never in the form of an individual or object.[18] Levinas explains how he was interested in participating in the conference because the paradigm of *vivir bien* – 'living well between ourselves', not 'living better' – coincides with his own notions of subject formation, in which 'subjectivity is not for itself, it is initially for another'.[19] I also hear him mentioning Enrique Dussel's writings on intersubjective ethics and the philosophy of liberation in Latin America, and I quickly take out my notebook and my pen.[20]

We buy traditional Nicaraguan pastries and juice made of local fruits from a woman who tries to squeeze down the aisle. The bus drives by the marketplace, and then curves out from the city centre. Derrida, Levinas and Spivak peek out of the bus windows in curiosity to see what the local scenery looks like. My enjoyment of this exotic atmosphere is disrupted when I recall Jamaica Kincaid's[21] harsh critique of the tourist's romanticizing gaze and Mary-Louise Pratt's[22] analysis of travel-writing as producing 'the rest of the world' for a European readership. I hear my companions sharing a sense of worry about the children in the streets begging for money and food. Romantic *bachata* rhythms play at a high volume and air blows in from the open windows. It becomes difficult to hear what the others are saying.

I cannot help staring at our guests, who continue their joyful yatter. Jacques Derrida seems to be the magnet that has drawn the three of them together. I read from my earlier notes that Derrida's critical engagement with Levinas provided Spivak with tools to 'to deconstruct the colonial legacy of the anthropological paradigm and to formulate the conditions of possibility for an ethical dialogue with the subaltern'.[23] Spivak, the translator of Derrida's *Of Grammatology* and also an advocate of deconstruction, is one of the few intellectuals putting into practice the suggestions made by the post-Enlightenment ethical movement associated with Derrida and Levinas.[24] However, despite the obvious interconnections and intersubjectivities among these writers, Spivak has chosen to engage in a direct dialogue with Derrida only. One of the possible reasons for this is Spivak's irritation over Levinas' gender-biased and Eurocentric writings on 'the other'.[25]

What Spivak, Derrida and Levinas all agree on is the impossibility of escaping epistemic violence when engaging with the other in discourse.[26] The upshot of this is that our journey, like others, is doomed to be unethical and to do violence to the ones I speak for and speak about, albeit in fictional terms. However, I am afraid that dwelling eternally on the problem of representation might bore my co-travellers – on the bus as well as on the page – and even lead to the entire journey being interrupted. Additionally, it would be rather contradictory to aim for perfect preparation and planning before an adventure. It would not be an adventure then, would it? So I just let the journey go on. A popular song by *Los Toros Band* plays on the radio, 'si tú estuvieras, sé que mi mundo sería diferente ...'[27]

I feel grateful that Spivak accepted the invitation letter that I sent her. Although the travel industry is to a large extent based on representation and utilization of the other, Spivak has been a rare guest in tourism discussions.[28] I must admit that while postcolonial analysis requires acknowledgement of the specific colonial history of the region,[29] Spivak is by no means a historian in Latin American Studies. However, on this trip she will most of all remind us of how our interactions with, and representations of, 'the subaltern other in the Third World'[30] are inevitably loaded by our positioning as researchers, experts or activists.[31] I remember now that in her view we should not understate the difficulty or importance of this fact, because even Foucault and Deleuze, 'the best prophets of heterogeneity and the other', are guilty of representing themselves as transparent.[32] I reckon Spivak will encourage us to acknowledge what social and institutional power relationships these representations establish or neglect. Put differently, what might the socio-political and ethical implications be of even those representations that seek to empower the local?[33]

There are in fact several reasons why I find Spivak's arrival to be a big surprise. Not least is her rejection of the persistence of anthropological fieldwork in which development practitioners, academics and students attempt to establish a fair dialogue with the 'local communities'. Additionally, she has directed untiring criticism toward the ways in which Northern, and local elite, researchers further their personal and institutional interests through field visits and data collection.[34] However, on our imaginary trip I have the opportunity not to reject, but to postpone responding to, this fundamental critique until I write my next research plan. Instead, today, I choose to focus on Spivak's warnings that we unwillingly reconstruct otherness and naturalize Western superiority and subjectivity through academic discussions.[35] I believe that, for now, this notion alone can make us painfully enough aware of the intractable controversies in the tourism and development settings that we are participating in at the moment.

The driver turns down the volume of the radio. Our special guests seem to have put the possible tensions between them on hold. Spivak and Levinas continue to lament the way in which oppressive dualist ontologies tend to prioritize the freedom of being over the relation with the other. They agree that conceiving the self, or being, as totalized and self-same almost inevitably either excludes or assimilates otherness.[36] 'Transcendence and goodness is produced as pluralism',

I hear Levinas exclaiming.[37] I think about tourist guide books or shiny travel magazines and their ways of reconstructing the individual freedom of a sovereign subject. However, perhaps I should not give any special award to the tourism industry for celebrating and always proceeding from the subject. In fact, I recall Olli Pyyhtinen explaining how, sociology included, 'there has been hardly any place left for the relational mode of the social', for co-existence and 'with'.[38] That is what he and also Jennie Germann Molz, Soile Veijola and Alexander Grit discuss in their chapters in our book: being-with has endless forms and potentialities.

The bus stops and the driver turns toward us to say: 'Bienvenidos a San Ramón'. As we pick our bags off the shelves, I catch myself silently repeating the word 'welcome'. It tastes different, more nuanced, now.

Walk on the coffee trails

In no time we receive the next welcome. Two local tourist guides, a young woman and a young man, have come to pick us up at the bus stop. They are wearing green t-shirts with the logo of *RENITURAL*, which stands for *la Red Nicaragüense de Turismo Rural Comunitario* (Nicaraguan Network for Rural Community-Based Tourism). During our walk to their home community we hear the entire story of tourism's arrival, starting from the Sandinista revolution in 1979. The guides tell us how in the 1980s, during the Contra war, the international solidarity movement brought the first foreign visitors to the area. At that point, tourism had not yet been organized and the visitors were 'attended as friends – not as tourists'. The first guests brought their own food and stayed with the local families for free. They expressed an interest in helping but also in learning about the collective spirit of the Nicaraguan socialist revolution and the newly founded coffee cooperatives.[39] These types of visits and different forms of unofficial help ended when the Sandinistas lost the election in 1990.[40]

On our way, we pass grazing horses and cows. A motorcycle passes us and our guides wave to the driver. The guides go on with their story about the global coffee price crisis around the year 2000, which had a severe impact on the cooperatives of small coffee producers. At that time, the regional and local coffee cooperative unions introduced an idea for a tourism programme. In addition to supplementary income and new contacts with coffee consumers, tourism was expected to contribute to gender equality and the creation of new job opportunities for young people. Since then, representatives of many bilateral aid organizations and non-governmental organizations (NGOs), as well as students and researchers, have become frequent visitors to the area.[41] In Spanish the initiative came to be called *Agro-Ecoturismo Comunitario,* while English-speaking visitors preferred to call it *Fair Trade Coffee Trail.*[42]

We hear how women and young guides, the ones committed to the new programme, participated in different forms of training in order 'to be able to receive guests'. Although many families and guides felt nervous and awkward with the first 'official' tourists, new contacts, positive experiences and better understanding of tourism activities helped the hosts gain confidence and enjoy the travellers' visits. In the early years of tourism, the guests stayed and slept in the same rooms as their host families. Our guides explain that while even today tourists stay with local families, there are now special rooms built for the purpose.[43] Some of the aid organizations and cooperative unions advised families to take relatively big loans, called microcredits, in order to pay for the construction of these rooms. When the number of visitors recently declined, the families ended up paying back the loans, and the interest, with coffee beans.

The refreshing walk through the picturesque countryside landscape – it could be straight out of the Nicaraguan edition of Lonely Planet – takes us to our final destination. It is a village of about 40 houses, an elementary school, two kiosks and a football field. There is also plenty of tourist signage, which makes the community more visitable.[44] The printed and painted boards welcome visitors to the community; indicate the houses that have tourist accommodation; tell how to get to the waterfall, the old gold mine and the scenic lookouts; help identify the trees; explain which coffee plants are organic; and provide reminders of generous donations from different aid organizations. We follow the guides and the signposts around the coffee plantations and get to know about 'coffee rust', the recent disease that has seriously damaged the plants. Many of us are shocked to hear how it will most probably take up to four years before coffee production can be restored to its earlier levels.

On doña Hilda's patio

At the end of the coffee tour, the guides take us to doña Hilda's house, where she has prepared us a typical Nicaraguan lunch: *gallo pinto* (rice and beans cooked separately and then fried together), white cheese, chicken, cabbage salad and corn tortilla bread. Actually, chicken is not part of the everyday menu but is saved for special occasions like this. Doña Hilda sits down to eat with us. She explains proudly how she has been providing accommodation for guests since the 1980s and how pleasant the experiences have been that her family has had with visitors from all over the world.

After finishing the meal and having a nice conversation on friendships in tourism settings, doña Hilda invites us to her patio to rest and enjoy the sounds of the forest that skirts her backyard. She encourages us to pay attention especially to the growling of the monkeys up in the mountains, knowing that many visitors consider the monkeys to be something very exotic. We gather more green plastic chairs onto the patio, and some of us find comfortable seating in a hammock. Gabriela asks doña Hilda about the current situation of tourism development. We can hear her voice become graver. After mentioning that there have been fewer tourists coming recently, doña Hilda expresses her concern about tourism projects completely changing her home and home village. She states that her family and friends have recently become more uneasy about the constant flow of visiting experts and consultants pointing out what has to be changed and improved in order to attract visitors. Doña Hilda tells about a recent visit by a specialist from a tourism programme called 'Modernize':

> This consultant came from the capital city, Managua. She looked at the rooms and said that we could not receive visitors in rooms like these. So she wanted to make changes in the place. She said we should have curtains, raise the ceilings and so on. We thought that we do not want to do this. It is too risky to take new loans for tourism development. This was something very strange to us. It seems to me that she wanted to change what rural community-based tourism is to make it like tourism in the cities. Honestly, it left us sad and offended.[45]

Indeed, it has been exactly these kinds of stories that have troubled me the most since my last visit here. I try to read the faces of the philosophers and their reactions to doña Hilda's anxiety. What do they think about the encounters where the guests feel obligated to teach their hosts how to receive visitors in their own homes? What do they think of visitors encouraging others to commoditize domestic forms of hospitality? I wonder whether doña Hilda chose to tell about these 'modernization efforts' in case we – as recently arrived guests – were also eager to find out what holes there were in the fabric of her hospitality.

Levinas nods and mumbles something about the silenced 'welcome of the other'. Then everybody remains quiet, as if waiting for doña Hilda to continue. She notices our troubled expressions and hastens to stress how this event does not reflect the relationships with all visitors:

> We think that this kind of tourism is something where we want to offer the visitor the best we have – our friendship and kindness. Before, the tourists did not have their own special room and they still liked this experience. They come here to learn about coffee production, to enjoy the nature and peace and to exchange and share experiences. The tourists normally know better what this kind of tourism is about.[46]

Doña Hilda's remark inspires us to exchange views on how we understand the purpose and goals of rural tourism – or of tourism in general. Most of us agree that the well-being of local families should be the starting point and tourism perceived as one of the possible instruments to promote it.[47] We also find a consensus that it can be misleading to expect that there are universal forms of sustainable or responsible tourism that could be replicated regardless of the local context, or regardless of the visitors. Claudia summarizes well how there is a troubling paradox at work here: despite the principle of local participation, the development intermediaries tend to conclude that the locals are not participating in the right way. In other words, ironically, the intended beneficiaries are blamed for being passive or steering the projects in wrong directions.[48] Hence, Claudia would like to hear how the philosophers might understand the foundation of participation. Doña Hilda rises up and apologizes that she has to step out for a moment. I see her disappearing into the kitchen, going backstage or to the 'back region', as we say in tourism studies.[49] Idania follows her to get a glass of water. As a matter of fact, in this kind of home-based tourism, the visitors have their own backstage passes.[50]

Spivak, who has a special insight into rural settings where the intermediaries are claiming to help the other – the so called subaltern with supposedly limited access to social mobility, leans forward in her chair. She prompts us to question the idea that the discourses of participation are somehow 'natural, good or incontestable'.[51] This implies that the promotion of participation always includes questions such as 'Why is something being developed in the first place?' and 'By whom?' And at the moment you feel it is really working, you have to stop and ask yourself 'For whom is it working?'[52] When listening to Spivak's questions, I wonder whether the recent celebration of local participation, indigenous knowledge and alternative models of development are actually blocking our view of a global vision and of macro levels of power in tourism encounters. In other words, too myopic a view of the local can cause us to overlook the structures that operate at the global level to reproduce inequality.[53] Spivak continues, stating that neglecting the legacy of colonialism and these kinds of modern forms of economic and political exploitation are epistemic violence, which forms a pertinent part of neoliberal orthodoxy. Derrida seems to argue with Spivak's contention that unawareness or ignorance of inequalities allows the maintenance of European 'theatres of responsibility'.[54] In other words, even emancipatory intentions in developing tourism cannot escape epistemic violence – and, unfortunately, community-based tourism projects easily create only new scenes and dramas in the theatre of responsibility.

Spivak explains further that one valid basis for discussion is to try to expose our blind spots, as Westerners or Nicaraguans, by challenging the following two assumptions. The first is that the political desire of the 'oppressed' and the political interests of development experts are identical. That is, we should question the naïve and paternalistic belief that *the other* is always willing to speak and participate, and capable of doing so, if *I* only listen and want her to participate. The second assumption is that the voices of the other can be recovered from the outside and that scholars can represent these voices as objective intermediaries. Accordingly, Spivak poses her famous, inevitable question, 'Can the subaltern speak?'[55] As she explains to us, the problem lies not in the inability of the other to speak but rather in the unwillingness and incapability of the 'culturally dominant' to listen. More precisely, Spivak takes the view that the privileged position which academic researchers and development consultants, for example, occupy, is the reason why the 'locals' cannot be heard. Instead, the other is always already interpreted. In other words, 'elite or hegemonic discourses are deaf to the subaltern, even when she or he speaks or resists'.[56] The consequence is what Spivak calls 'epistemic ignorance' or 'epistemic violence': the trivialization and invalidation of ways of knowing that fall outside of the West's, and the local elite's, languages, epistemic traditions and philosophies.[57]

Out of the blue a hen hops into the hammock to accompany Elba, Danilo and Mónica. The surprise guest almost makes them fall onto the floor with laughter.

The challenge of unlearning

To me it sounds rather dramatic, dubious, or at least rather materialistic, to divide subjects to subalterns and privileged ones. I wonder if I have understood Spivak's approach correctly. Possibly there are unlimited ways of understanding it. In any event, my attention is drawn to the significance of her ideas in tourism schemes. Hence, I ask her whether she sees the privileged position of guests as perhaps undermining the potential of promoting ethical forms of visiting and travelling. In that case, could it be that the pre-constructed categories of what tourism is and how it should help local communities most probably, unintentionally, restricts the subjectivities, choices and voices of the other? And how do the tensions between the West and the Rest or the core and the periphery lead to our imposing top-down definitions of development and sustainability in tourism?

Spivak points out how she is unfamiliar with tourism projects but confirms that the privileged position of the self restricts the possibility of listening to the other. For that reason she challenges us, and herself, to *unlearn our privileges as loss*. Spivak tells us that the task requires, firstly, noticing that one's class, race, gender, ethnicity, nationality, ideology, education, occupation, language or even access to Internet can create relative advantage.[58] Next she asks us to think about what kind of prerogatives, we – as individuals and as a group – might have as visitors in this particular context. However, she also warns us of starting to sound pious and thereby only reinforcing our positions.[59] While exchanging views on this, we notice how our relative advantages as dominant groups can be based on many roles – on being a tourist, working for development, researching tourism and so on.[60] It becomes evident that simply travelling from 'the cores' to 'the peripheries' strengthens the subjectivity of the mobile traveller. Or as another captivating philosopher, Sara Ahmed, has expressed it, some of us are afforded agency within the global, by relegating others to 'local' spaces.[61]

Spivak goes on, clarifying that, secondly, 'unlearning one's privilege as loss' and changing one's mindset mean recognizing one's own prejudices and learned responses.[62] To put it in another way, our prerogatives have given us only limited knowledge and have prevented us from gaining new understanding. Therefore we are simply not equipped to understand different ways of knowing.[63] Spivak's approach means denying the idea of the Enlightenment whereby the world is expected to be knowable through observation. There are certain knowledges, experiences and existences that are closed off from the 'privileged view'.[64] I understand Spivak's thinking as suggesting that a desire to take into account, for instance, local knowledge about tourism is misleading and not enough. Notably, paternalistic intentions, unaware as they are of these limitations, can also be seen as a form of silencing, objectifying and trivializing the other.[65]

The same curious hen comes back to the patio, tilting its head from side to side; but this time the team occupying the hammock is more alert, and the hen gives up on its plan to join them. Spivak continues that, furthermore, in an ethical relationship with the other, it is not sufficient only to learn to unlearn, but one must 'learn to learn anew'. For her, these processes are central if one is to deconstruct the positivist and essentialist paradigms of representation that underpin the claims of many Western, or local elite, intellectuals to *speak for* and *speak about (re-present)* the 'oppressed'. Spivak argues that two dislocated subject perspectives, the political sense of *speaking for* the other, on the one hand, and the aesthetic/philosophical sense of *speaking about* the other, on the other, are the two common forms of representation.[66] Indeed, as Levinas interjects, interrupting Spivak gracefully, saying we must notice that this kind of relation between self and other resonates within the conventional '*said*' instead of '*the saying*'. Therefore, significantly, the ethical relations we discuss and explore are not metaphysical but take place in language, as traces of the ethical.[67]

A short pause in the conversation allows Elba to pose a question. 'But could ethical representation even be possible?' she asks. Levinas suggests to her that this would require desiring the omnipresent excess of the saying instead of privileging the said. Obviously taking a risk of oversimplifying the case, Levinas outlines how the said could be defined as something with totality and limits, while the saying refers to openness, to 'the living word', to the gesture toward another being.[68] Derrida suggests that we think about the two co-existing facets, the said and the saying, in connection with the conditional laws of hospitality and absolutely unconditional welcoming. Just like the law and the ethics of hospitality are not opposites or alternatives to each other, neither are the said and the saying.[69] Instead, the saying expands 'the potentially reductive and oppressive boundaries of the said'.[70] Spivak agrees, using the same concepts as Levinas, that this 'double session of representation' could be de-established by the missing (third) sense of *the ethical saying*.[71]

Levinas and Spivak continue with a debate on whether and how ethical saying could be achievable. Spivak argues that ethical singularity could be reached 'through painstaking labour', albeit labour that excludes fieldwork contexts.[72] Just when Levinas begins to explain in more detail why he thinks that ethics and responsibility transcend the individual agency of self, doña Hilda returns with a steaming coffee-pot.[73] We are getting, theories aside, freshly roasted local coffee. The interruption offers us a needed break to digest all that we have just heard. The coffee tastes good, as always; it is much sweeter and softer than back home. Yet not even delicious coffee can shake from my mind Spivak's critique of our overzealous willingness to represent the other. So, paradoxically, my knowledge about alternative forms of participatory tourism development can seriously hinder me from promoting these alternatives. I can see confused smiles among the other tourism scholars. Many of us must have been waiting to gain concrete information on how to 'represent the other' *in a better way* in project documents, academic research reports, tourism advertisements and so on. However, Spivak's contemplations have made us question whether we are actually needed here at all. Her proceeding by a negative[74] has also prompted scepticism whether ethical relations could ever flourish in tourism projects, especially when the projects are focused on developing, teaching and helping.

Reciprocities of teaching

While drinking doña Hilda's coffee, we try to recognize our own positions in tourism and development discourses. It is obvious that the latter are hegemonic discourses; the North maintains the position of superiority, agency and the role of an adult, while the child-like South is taught and helped to follow linear, Western-style development as a norm.[75] Or, to put it another way, the problems and challenges are found in rural communities, for instance, and the guests are the ones capable of providing the needed solutions. This mindset allows the subjects of the West to be conserved. But when I listen to my Nicaraguan colleagues, it appears that the same kind of (institutional) power relations exist also between cores and peripheries, urban and rural areas within the so-called Global South. Likewise, they exist within the so-called Global North, as Rauna Kuokkanen shows in her work on the ambiguity of hospitality and responsibility in academia toward indigenous people and indigenous epistemes.[76]

However, it is also a questionable privilege to be able to reject development or to become strictly 'anti-tourism' when having the access to material abundance oneself.[77] Having said this, I continue believing that tourism intermediaries *can* support rural communities in developing tourism services, creating new contacts and gaining the needed self-confidence.[78] As if reading my thoughts, Gabriela breaks the silence and asks what kind of advice the philosophers could give to lecturers and trainers in the field of hospitality, to those of us who are teaching tourism, hospitality and development in the classrooms and the field. She would like to understand how the process of unlearning is coupled with teaching. It is Levinas's turn to continue from here. His answer sounds reassuring, although a bit cryptic. In fact, I get a feeling that his prime interest is in neither teaching nor learning but in *being taught*. Levinas explains that welcoming teaching means *receiving more than I contain*. Therefore, welcoming teaching refers to openness and infinity – to a soul capable of containing more that it can draw from itself, to a soul that does not take its own interiority for the totality of being.[79] Derrida joins in and they state in concert that welcoming teaching is receiving from the other beyond the capacity of 'the I', which means precisely *to have the idea of infinity*.[80]

I interpret Levinas's notion to mean that if we have earlier perceived our roles as 'tidy guests' who could come to help and teach, now, all of a sudden, we should be able to adopt the role of learner. This means that we need to question the learned position of guiding the other. To me, Spivak's call for *unlearning* and Levinas's approach of *receiving more than I contain* sound, when put together, like an idea of mutual learning and teaching. In the context of rural tourism, it would require rejecting the assumption that teaching is directed only toward 'host' communities, that is, that the flow of teaching has only one significant direction.[81]

Hospitality as a tool for unlearning

In a timely manner, Danilo, one of the younger scholars, acknowledges doña Hilda's advice or suggestions to us as visitors. Instead of answering directly, doña Hilda refers to the earlier visit of 'the consultant from Managua':

> We have thought that we will accept it if visitors do not come. And in any case, we have now expressed to the local tourism coordinators that this particular consultant is no longer *welcome* here. We do not want to receive her here.[82]

At this, the faces of Levinas and Derrida light up. Derrida proposes that the notion of hospitality, that is, the 'welcome *of* the other', could be applied here as a valuable analytical tool. He continues that hospitality provides access to the possibility of ethics, to ethics *as* hospitality.[83] Levinas tells us that the obligation to welcome and to do justice to the other restricts the freedom of the self and that 'it calls in question the naïve rights of my powers'.[84] It means that the concept of subjectivity could be transformed and redefined through the pivotal thought of hospitality to refer to 'the subject of welcome'.[85] I imagine how this kind of subjectivity is not based on conditions – such as accepting and approving the hospitality of the other *only if they have proper curtains*. While listening to the discussion, I begin to recognize the potential of hospitality in the process of 'learning to unlearn' and 'learning anew', and learning anew the desire to be 'taught'.

We all nod to Levinas's idea of transforming ethical subjectivity: that the subjectivity and responsibility of the self is constituted by the 'welcome *of* the other'. Idania interprets the idea to mean 'if you behave badly during your visit, you are most probably not going to be invited again'. It is easy to agree with her. This sounds logical. However, Derrida reminds us that we should not reduce hospitality to refer to the physical invitation only. The ethical relation between the self and the other, according to Levinas, is not a plainly conscious choice made by a rational subject for the other.[86] Instead, goodness and ethics exist without egoism, and especially without the cognitive emprise of the ego that seeks to reduce all otherness to itself.[87] Levinas takes a sip of his coffee and goes on. He explains how, for him, responsibility for the other does not begin from the self's knowledge but from the call of the other. Responsibility is not mine, he insists. Therefore it must be understood as being first a welcome *of* the other, and then a welcome *to* the other as a response to the 'yes' of the other.[88]

Out on the patio, it sinks in that Levinas's and Derrida's thought on subjectivity is drastically different from the contemporary encounters in tourism, development aid and, lamentably, the academic world as well.[89] Today, communities in economically marginalized areas, for instance the home village of doña Hilda, tend to be perceived as the arenas of unconditional or unlimited hospitality, where subjects from the industrialized centres can carry out their supposedly moral work. Just to say 'yes' to the anticipated welcome of the local populations is often considered an admirable act as such.[90] However, in this process the local actors or, in Spivak´s words, 'the subaltern', rarely speak; their open invitation *to* the other to visit and intervene is taken for granted. It means that today's tourists, development workers and researchers presume that we have received the open welcome *of* the other – which defines our own subjectivity – while neglecting the subjectivity of the other. However, the imbalance, silencing and de-subjectifying of the other has been regularly dismissed even in the discourses on responsibility in tourism.

Levinasian thought denies the possibility of treating moral action as a free project of a spontaneous subject. Derrida confirms that respecting the priority of the 'yes' *of* the other over the 'yes' *to* the other would completely alter the approach to the question of decision and responsibility.[91] Adapting Levinas's mindset, hospitality could be used to de-establish *speaking for* and *speaking about* the other and to envision what Spivak and Levinas referred to previously as 'the ethical force of saying'. This ethical saying does not aim to define the other as an object of knowledge, but merely has a desire for infinity, openness and receptivity.

At the end of the day

We can hear the sound of a bus arriving in the village. In no time, curious youngsters come onto the patio to greet us. We change the subject to football, travelling and the temporary silence of the monkeys. I try to remember who it was that was cheering for Real Madrid and who for Barcelona, and I get to answer questions about my family back home. Despite the joyful moment, I cannot help sinking into a state of hyper-self-reflectivity, as Ilan Kapoor would express it. Although everything has gone better than well, I feel blameworthy for welcoming the philosophers to visit San Ramón and the home of doña Hilda. It makes me marvel how the open invitation of the other apparently became self-evident. Have all the tourism advertisements convinced us, the wealthy guests, to feel welcome to arrive whenever and wherever?[92] Who 'sold' this kind of an idea of a welcome to the academics, researchers and development practitioners? Can we explain our self-confidence as visitors through our possible self-identification as 'tidy guests' – neat guests with great skills for organizing the lives of the others?[93] However, guests who feel free to invite more visitors can also appear as rude guests who take over hosting the party. My only consolation is that today I did this only in writing.

Even so, I know now that our biased position as 'experts' in something can easily become a loss. Instead of turning my narrative into a mere 'introspection exercise' I redirect my attention to the doors that we have been opening, or the new doorways that have been making, in today's talks. During our stay we have been able to reflect on our restricted and narrow understandings of hospitality. Most of all, Levinas's and Derrida's ideas on hospitality have invited us to conceive subjectivities as a primordial ethical experience. Even if an unconditional welcome is 'the experience of the impossible', as Spivak[94] put it, it still represents something 'to aspire to and fall short of'. [95] I recall doña Hilda mentioning how, for her, hospitality means saying 'Welcome to my home' and hoping 'that the visitors feel like they are part of the family. And they feel at peace'.[96]

The guides return and we make plans for the rest of the afternoon. We decide to walk down to see the cacao plantations and to collect some vegetables for dinner. When walking with the guides, we talk eagerly with one another of ways of unlearning our fixed roles as loss, and how learning to learn anew means questioning our will to teach, develop, modernize, thematize, speak, theorize, define, correct, enlighten and so on. Following Jennie Germann Molz's ideas of unfinishedness in her chapter in this book, we try to think how to build unfinished design into the development projects. In these kinds of field sites, this policy would, paradoxically, mean that *failure* in planning, teaching, theorizing, developing, modernizing, colonizing, speaking and defining could be interpreted as *success*.[97] Success can be a failure to force someone to buy new curtains or giving up on intentions of knowing, defining and totalizing the other.

The discussion continues later on, when doña Hilda tries to teach our guest philosophers how to make tortillas on an open fire, with varying results. Spivak laughs and encourages us to think about working without guarantees[98] and seeing one's failure as a success,[99] to which Derrida solemnly nods and cites himself: 'I must be unprepared, or prepared to be unprepared, for the unexpected ...'[100] At the dinner table, eating the misshapen but nevertheless tasty tortillas, we try to imagine the arrival of a visitor without project documents and authority – a visitor with lighter gear, appreciation toward hospitality and money to pay for the local services of lodging and food. We imagine a guest with a desire to be taught by the subject of the welcome.

Notes

1. For examples of various challenges, see: Laurence Chalip and Carla A. Costa, 'Clashing Worldviews: Sources of Disappointment in Rural Hospitality and Tourism Development', Hospitality & Society, 2.1,

2012, pp. 25–47; Maria José Zapata, Michael C. Hall, Patricia Lindo and Mieke Vanderschaeghe, 'Can Community-Based Tourism Contribute to Development and Poverty Alleviation? Lessons from Nicaragua', in Jarkko Saarinen, Christian M. Rogerson and Haretsebe Manwa (eds), Tourism and the Millennium Development Goals: Tourism, Local Communities and Development, London, Routledge, 2012, pp. 98–122; Ernest Cañada, 'Perspectivas del Turismo Comunitario: Cómo Mantener Vivas las Comunidades Rurales', http://blog.pucp.edu.pe/item/93900/mundo-perspectivas-del-turismo-comunitario-como-mantener-vivas-las-comunidades-rurales, website accessed on 22 October 2011.

2. I have collected my ethnographic data for my PhD thesis in tourism studies during the years 2008–2013. Since 2011, I have been working in the research project *POLITOUR: Policies and Practices of Tourism Industry – A Comparative and Interdisciplinary Study on Central America* (Academy of Finland, 138589), led by Jussi Pakkasvirta at University of Helsinki. The principal supervisor of my thesis is Soile Veijola.

3. For inspiration of doing togetherness differently, see: Jennie Germann Molz and Sara Gibson (eds), *Mobilizing Hospitality. The Ethics of Social Relations in a Mobile World*, Aldershot, Ashgate, 2007; and Sara Ahmed, *Strange Encounters. Embodied Others in Post-Coloniality*, London, Routledge, 2000.

4. See for instance: David Bell, 'Tourism and Hospitality', in Tazim Jamal and Mike Robinson (eds), *The SAGE Handbook of Tourism Studies*, London, SAGE Publications Ltd., 2009, pp. 19–34; Germann Molz and Gibson, *Mobilizing Hospitality*.

5. Emmanuel Levinas, *Totality and Infinity: An Essay on Exteriority*, translated by A. Lingis, Pittsburgh, Duquesne University Press, 1969; Jacques Derrida, *Adieu to Emmanuel Levinas*, translated by P. A. Brault and Michael Naas, Stanford, CA, Stanford University Press, 1999. The preface of the book is Derrida's funeral oration for Levinas on 27 December 1995, while the second part, titled 'The Word of Welcome' reproduces Derrida's opening lecture at the 'Homage to Emmanuel Levinas' conference one year later.

6. Jacques Derrida, 'The Principle of Hospitality', *Parallax*, 11.1, 2005, pp. 6–9.

7. Ibid.; Jacques Derrida, *Adieu to Emmanuel Levinas*, Paris, Galilée, 1999, p. 19.

8. Soile Veijola and Eeva Jokinen, 'The Body in Tourism', *Theory, Culture & Society*, 11.3, 1994, pp. 125–151.

9. Gayatri Chakravorty Spivak, 'Can the Subaltern Speak?' in Cary Nelson and Lawrence Grossberg (eds), *Marxism and Interpretation of Culture*, Chicago, University of Illinois Press, 1988, pp. 271–313.

10. Gayatri Chakravorty Spivak, *A Critique of Postcolonial Reason: Toward a History of the Vanishing Present*, Cambridge, MA, Harvard University Press, 1999, pp. 308–309; Leerom Medovoi, Raman Shankar and Benjamin Robinson, 'Can the Subaltern Vote?' *Socialist Review*, 20.3, 1990, pp. 133–149.

11. See in particular Mick Smith, 'Ethical Perspectives: Exploring the Ethical Landscape of Tourism', in Tazim Jamal and Mike Robinson (eds), *The SAGE Handbook of Tourism Studies*, London, SAGE, 2009, pp. 623–626; Germann Molz and Gibson, *Mobilizing Hospitality*.

12. Derrida, *Adieu to Emmanuel Levinas*, p. 20; For a comprehensive treatment of hospitality in International Relations, see: Gideon Baker (ed.), *Hospitality and World Politics*, New York, Palgrave Macmillan, 2013.

13. Derrida, *Adieu to Emmanuel Levinas*.

14. Ibid., p. 48; For the articulation and discussion of Derrida's treatment of the motif of subjectivity in *Adieu á Emmannuel Levinas*, Paris, Galilée, 1997, see François Raffoul, 'The Subject of the Welcome. On Jacques Derrida's Adíeu á Emmanuel Levinas', in *Symposium*, 2.2, 1988, p. 214.

15. Ibid., p. 219; Derrida, *Adieu to Emmanuel Levinas*, p. 25.

16. Smith, 'Ethical Perspectives', pp. 613–630; for examples of exceptions, see: Cara Aitchinson, 'Theorizing Other Discourses of Tourism, Gender and Culture. Can the Subaltern Speak in (Tourism)?' *Tourism Studies*, 1.2., 2001, pp. 133–147; Jo Ankor and Stephen Wearing, 'Gaze, Encounter and Philosophies of Otherness', in Omar Moufakkir and Yvette Reisinger (eds), *The Host Gaze in Global Tourism*, Oxfordshire, CABI, 2012, pp. 179–191.

17. Jon Bergeå, *Class Travellers in Chicken Buses, Backpacking, Economic Inequality and Cultural Identity in Guatemala*, Minor Field Study 2002, Peace and Development Research Institute, Gothenburg University, http://www.resurs.folkbildning.net/PageFiles/10619/class%20travellers. PDF, website accessed on 10 November 2013.

18. Arturo Escobar, 'Latin America at a Crossroads', *Cultural Studies*, 24.1, 2010, pp. 1–65; Eija Ranta-Owusu, 'Governing Pluralities in the Making. Indigenous Knowledge and the Question of Sovereignty in Contemporary Bolivia', *Journal of the Finnish Anthropological Society*, 35.3, pp. 28–48, 2010; 'Plan Nacional de Desarrollo'. Bolivia Digna, Soberana, Productiva y Democrática para Vivir Bien. Lineamientos Estratégicos 2006–2011. República de Bolivia. Ministerio de Planificación del Desarrollo, La Paz – Bolivia, Septiembre, 2007, http://www.ine.gob.bo/indicadoresddhh/archivos/Plan%20Nacional%20de%20Desarrollo.pdf, website accessed on 20 January 2013; For more discussion on indigenous ontologies and epistemologies, see Rauna Kuokkanen, *Reshaping the University. Responsibilities, Indigenous Epistemes, and the Logic of the Gift*, Vancouver, UBC Press, 2007, pp. 59–62.

19. 'Plan Nacional de Desarrollo', p. 8 quoted in Ranta-Owusu, 'Governing pluralities', p. 31; Tara Daly, 'The Intersubjective Ethic of Julieta Paredes' Poetic', *Bolivian Studies Journal, Revista de Estudios Bolivianos*, 15–17, 2010, pp. 237–263; Emmanuel Levinas, *Ethics and Infinity*, Pittsburgh, Duquesne University Press, 1982, p. 96.

20. Enrique Dussel, *Twenty Theses on Politics*, translated by G. Ciccariello-Maher, Duke University Press, Durham, 2008; Moraña, M., Dussel, E. and Jáuregui, C. A. (eds), *Coloniality at Large: Latin America and the Postcolonial*

Debate, London, Duke University Press, 2008; John E. Drabinski, *Levinas and the Postcolonial: Race, Nation, Other*, Edinburgh, Edinburgh University Press, 2011, pp. 3–8.

21. Jamaica Kincaid, *A Small Place*, New York, Farrar, Strauss & Giroux, 1988.
22. Mary Louise Pratt, *Imperial Eyes: Travel Writing and Transculturation*, London and New York, Routledge, 1992.
23. Stephen Morton, *Gayatri Spivak: Ethics, Subalternity and the Critique of the Postcolonial Reason*, Cambridge, MA, Polity Press, 2007, p. 61.
24. Jane Hiddleston, *Understanding Postcolonialism*, Stocksfield, Acumen Publishing Limited, 2009; Zuzanna Ladyga, 'Tracing Levinas in Gayatri Spivak's Ethical Representations of Subalternity', in Ewa B. Luczak, Justyna Wierzchowska and Joanna Ziarkowska (eds), *In Other Words. Dialogizing Postcoloniality, Race, and Ethnicity*, Warzava, Encounters, 2012, pp. 221–230.
25. Gayatri Chakravorty Spivak, *Outside in the Teaching Machine*, New York, Routledge, 1993, pp. 166–167; Ladyga, 'Tracing Levinas'.
26. Jacques Derrida, *Of Grammatology*, translated by G. C. Spivak, Baltimore, Johns Hopkins University Press, 1976; Spivak, 'Can the Subaltern Speak?'; Emmanuel Levinas, *Otherwise than Being, or Beyond Essence*, translated by A. Lingis, Pittsburgh, Duquesne University Press, 1998.
27. 'If you were, I know that my life would be different ...'
28. For examples of inviting Spivak to tourism studies, see Aitchinson, 'Theorizing Other Discourses of Tourism'; Keith Hollinshead, 'Playing with the Past: Heritage Tourism Under the Tyrannies of Postmodern Discourse', in Chris Ryan (ed.), *The Tourist Experience*, London, Continuum, 2002, p. 190.
29. Teivo Teivainen, *Pedagogía del poder mundial*, INTO eBooks, 2004, p. 10; Moraña et al., *Coloniality at Large*, p. 5.
30. I understand Spivak's way of using these controversial concepts of 'subaltern' and 'Third World' as strategic essentialism. At the same time, she encourages us to engage in a persistent critique of these kinds of hegemonic representations.
31. Cheryl McEwan, *Postcolonialism and Development*, New York, Routledge, 2009, p. 275; Dianne Dredge and Rob Hales, 'Community Case Study Research', in Larry Dwyer, Alison Gill and Neelu Seetaram (eds), *Handbook of Research Methods in Tourism: Quantitative and Qualitative Approaches*, Cheltenham and Northampton, MA, Edward Elgar Publishing, 2012, p. 419.
32. Spivak, 'Can the Subaltern Speak?' pp. 275–283.
33. Spivak, *A Critique of Postcolonial Reason*, p. 284; For an excellent analysis about the relevance of Spivak's work in the field of development, see: Ilan Kapoor, 'Hyper Self-Reflexive Development? Spivak on Representing Third World "Other"', *Third World Quarterly*, 25.4, 2004, p. 628.
34. Ibid., pp. 632–633; Jenny Sharpe and Gayatri Chakravorty Spivak, 'A Conversation with Gayatri Chakravorty Spivak: Politics and the Imagination', *Signs*, 28.2, 2003, pp. 619–620; Spivak, 'Can the Subaltern Speak?' p. 275.

35. Ibid., p. 271; Kuokkanen, *Reshaping the University*, p. 6.
36. Drabinski, *Levinas and the Postcolonial*, pp. 50–53; Levinas, *Totality and Infinity*, pp. 45–46.
37. Ibid., p. 304
38. Olli Pyyhtinen, 'Being-with: Georg Simmel's Sociology of Association', *Theory, Culture & Society*, 26.108, 2009, p. 121.
39. Emily Höckert, 'Community-Based Tourism in Nicaragua: A Socio-Cultural Perspective', *The Finnish Journal of Tourism Research*, 7.2, 2011, pp. 14–15.
40. See Medovoi, et al., 'Can the Subaltern Vote?'
41. Cañada, 'Perspectivas del'; Fransisco J. Pérez, Oscar D. Barrera, Ana V. Peláez and Gema Lorío, *Turismo Rural Comunitario como Alternativa de reducción de la pobreza rural en Centroamérica*, Managua, EDITASA, 2010, pp. 56–58; Höckert 'Community-Based Tourism'; Zapata et al. 'Can Community-Based Tourism'.
42. UCA San Ramón, http://www.tourism.ucasanramon.com.website accessed on 10 October 2013.
43. Höckert, 'Community-Based Tourism'.
44. See Bella Dicks, *Culture on Display: The Production of Contemporary Visitability*, Maidenhead, McGraw Hill, 2004.
45. Direct quotation from an interview with a woman providing accommodation for tourists, conducted by the author in May 2013.
46. Ibid.
47. Cañada, 'Perspectivas del'; Pérez et al. *Turismo Rural Comunitario*; Höckert 'Community-Based Tourism'.
48. See for example: Maria Eriksson Baaz, *The Paternalism of Partnership. A Postcolonial Reading of Identity in Development Aid*, London, Zed Books Ltd., 2005; and Chalip and Costa, 'Clashing Worldviews'.
49. Dean MacCannell, *The Tourist: A New Theory of the Leisure Class*, London, The Macmillan, 1976, pp. 94–95.
50. Erika Andersson Cederholm and Johan Hultman, 'Bed, Breakfast and Friendship: Intimacy and Distance in Small-Scale Hospitality Businesses', *Culture Unbound: Journal of Current Cultural Research*, 2, 2010, pp. 365–380.
51. Kapoor, 'Hyper Self-Reflexive Development?', p. 642.
52. Ibid.; Sharpe and Spivak, 'A Conversation with', p. 623.
53. For excellent discussions on micro and macro levels of inequalities, see in particular: Kuokkanen, *Reshaping the University*, pp. 109–112; Wanda Vrasti, *Volunteer Tourism in the Global South: Giving Back in Neoliberal Times*, London and New York, Routledge, 2013.
54. Spivak, 'Can the Subaltern Speak?'; Morton, *Gayatri Spivak*, p. 60.
55. Ibid; Kapoor, 'Hyper Self-Reflexive Development?'
56. Ibid, p. 639; Spivak, 'Can the Subaltern Speak?' p. 308; for another example from Nicaragua, see Anja Nygren, 'Local Knoledge in the Environment-Development Discourse: From Dichotomies to Situated Knowledges', *Critique of Antropology*, 19.3, 1999, pp. 267–288.

57. Kuokkanen, *Reshaping the University*, pp. 66–68; Sharpe and Spivak, 'A Conversation with', p. 613.
58. Ibid., pp. 617–618; Spivak, 'Can the Subaltern Speak?'; for discussion on 'white bodies unlearning their privilege', see Vrasti, *Volunteer Tourism in the Global South*, pp. 123–124.
59. Sharpe and Spivak, 'A Conversation with', p. 620.
60. For analysis of identity and privilege in development aid, see: Kapoor, 'Hyper Self-Reflexive Development?'; Eriksson Baaz, *Paternalism of Partnerhip*, p. 106; for analysis of tourism, see Bergeå, *Class Travellers in Chicken Buses*; Vrasti, *Volunteer Tourism in the Global South*, p. 22.
61. Ahmed, *Strange Encounters*, p. 172; I want to thank my colleague Piia Lavila at University of Helsinki for the discussion about this thought.
62. Spivak, *The Postcolonial Critic*, p. 20; McEwan, *Postcolonialism and Development*, p. 68.
63. Kuokkanen, *Reshaping the University*, pp. 3, 103.
64. McEwan, *Postcolonialism and Development*, p. 68.
65. Spivak, 'Can the Subaltern Speak?' pp. 275–276.
66. Ibid.
67. Ladyga, 'Tracing Levinas', p. 227; Levinas, *Otherwise than Being*.
68. Ibid.; See also Daly, 'The Intersubjective', pp. 237–243.
69. Derrida, *Adieu to Emmanuel Levinas*.
70. Levinas, *Otherwise than Being*, p. 44, quoted in Hiddleston, *Understanding Postcolonialism*, p. 19.
71. Ladyga, 'Tracing Levinas'; Gayatri Spivak, 'Translator's Preface and Afterword', in Mahasweta Devi, *Imaginary Maps: Three Stories*, London: Routledge, 1995, p. xxiv.
72. Ibid., p. xxiv; Spivak, 'Can the Subaltern Speak?' pp. 275–276.
73. Levinas, *Totality and Infinity*.
74. Kapoor, 'Hyper Self-Reflexive Development?' p. 639
75. Eriksson Baaz, *The Paternalism of Partnership*, pp. 39–42; Arturo Escobar, *Encountering Development: The Making and Unmaking of the Third World*, Princeton, Princeton University Press, 2012; Teivainen, *Pedagogía del poder mundial*, pp. 6–8.
76. Kuokkanen, *Reshaping the University*.
77. Eriksson Baaz, *The Paternalism of Partnership*, pp. 163–164.
78. Höckert, 'Community-Based Tourism'.
79. Anna Strhan, *Levinas, Subjectivity, Education. Towards an Ethics of Radical Responsibility*, West Sussex, Wiley Blackwell, 2012; Levinas, *Totality and Infinity*, p. 180.
80. Ibid., p. 51; Derrida, *Adieu to Emmanuel Levinas*, pp. 18, 27.
81. Paolo Freire, *Pedagogy of Freedom: Ethics, Democracy and Civic Courage*, Lanham, MD, Rowman and Littlefield, 1998; Sharpe and Spivak, 'A Conversation with', p. 620; Teivainen, *Pedagogía del poder mundial*; Dredge and Hales, 'Community Case Study Research'.
82. Direct quotation from an interview with a woman accommodating tourists, conducted by the author in May 2013.

83. Raffoul, 'The Subject of the Welcome', p. 214.
84. Levinas, *Totality and Infinity*, p. 84.
85. Ibid. p. 299; Derrida, *Adieu to Emmanuel Levinas*, p. 54; See also Raffoul, 'The Subject of the Welcome'.
86. Levinas, *Ethics and Infinity*.
87. Simon Critchley 1999, *The Ethics of Deconstruction. Derrida and Levinas*, Edinburgh, Edinburgh University Press, pp. 4–5; Levinas, *Totality and Infinity*, pp. 305–306.
88. Ibid., p. 93; Levinas, *Otherwise than Being*; Derrida, *Adieu to Emmanuel Levinas*, pp. 22–25.
89. Strhan, *Levinas, Subjectivity*; Kuokkanen, *Reshaping the University*, pp. 128–142.
90. Vrasti, *Volunteer Tourism in the Global South*, pp. 112–114.
91. Derrida, *Adieu to Emmanuel Levinas*, pp. 23, 53–55, quoted in Raffoul, The Subject of the Welcome', p. 215.
92. I thank Jennie Germann Molz for this thought.
93. I want to thank Alexander Grit for helping me to develop this thought.
94. Spivak, 'Translator's Preface', p. xxv.
95. Derrida, 'The Principle of Hospitality'; Baker, *Hospitality and World Politics*, p. 3; Germann Molz and Gibson, *Mobilizing Hospitality*, p. 5.
96. Direct quotation from an interview with a woman accommodating tourists, conducted by the author in May 2013.
97. Kapoor, 'Hyper Self-Reflexive Development?' pp. 641–642.
98. Gayatri Chakravorty Spivak, 'A Note on the New International', *Parallax*, 7.3, 2001, p. 15.
99. See Kapoor, 'Hyper Self-Reflexive Development?' p. 644.
100. Quoted in Richard Kearney and Mark Dooley (eds), *Questioning Ethics*, London and New York, Routledge, 1999, p. 70.

6

Messing around with Serendipities

Alexander Grit

Introduction

Today is perhaps not the best day to travel into philosophy – nor the Zuiderzee museum in Enkhuizen for that matter, since the weather is letting us down. It is November in the Netherlands, and the rain is pouring down. We park the car and my ten-year-old daughter complains about her sandwiches. I used the wrong peanut butter this morning. Which is not a good start. I admit my mistake and hope that somehow her attention shifts to something else soon. My friend Kwame joins us at the gates of the museum, and we walk to the big signs located in the front of the open air museum. I read them while my daughter reminds me that such mistakes as peanut butter are very serious and should not be taken lightly. My friend Kwame agrees to this with a wink to my daughter, and I realize that this partnership may affect my day. While I read, my glove falls on the wet ground. I will tell you quickly what's on the sign. Following a large flood in the crisis year of 1930, the Dutch government decided to build a 30-kilometer dam to close the Zuiderzee and turn the sea into a large lake. A range of local settlements such as fishing villages changed dramatically. People lost their jobs and the once proud sea harbours became lakeside villages. I look at my friend Kwame to see what he will say about this change, but he has this peanut butter pact with my daughter just to tease me. I know that the French philosophers Gilles Deleuze and Felix Guattari would call this a huge *de-territorialization* whereby a *life world* falls apart and moves away from a hierarchical and rigid context that imposes

122

singular identities and meanings.[1] Kwame and I are both Deleuzian and Guattarian scholars, and I hear Kwame say: 'I see a process of de-territorialization whereby a well-known familiar "area" disintegrates'. *Territorialization* in the other hand is a process whereby rules, meanings and fixed ideas are imposed on an 'area'. For example the process of Christianization of Northern Europe during the middle ages can be seen as a territorialization whereby one belief system was rather forcefully introduced. It is interesting to see how the process of de-territorialization happens and how *re*-territorialization takes place.[2]

The Zuiderzee museum is a place where old buildings re-territorialize, so these houses can form rather known territory in a new context. I do not need my friend Kwame for that; he, by the way, is joking with my daughter and excludes me, just because of that sandwich. And yes, I recognize this process of de-territorialization which does not re-territorialize in the previous, original configuration as I read: 'To preserve parts of this culture [daily life around the Zuiderzee], buildings were taken apart, collected, rebuilt and conserved in the Zuiderzee Museum in Enkhuizen'.[3] The Museum Park was completed in 1983, after years of preparation. The museum provides an image of how people used to live and work around the Zuiderzee between 1880 and 1930.[4]

My friend Kwame also starts to read the sign. I know already what he will say. He will lean on Deleuzean terminology. And, yes, I am right. He straightens his back and formally states: 'What I see is a process of de-territorialization, initiated by the line of flight by building a dike [Afsluitdijk] and whereby the parts re-territorialize in museum space'.

I briefly nod and notice the smell of peanut butter. He probably ate my daughter's sandwich. In this respect Kwame has become both a philosopher of a utilitarian stripe and a parasite as well. (On parasites, see the chapter by Olli Pyyhtinen in this book.) My friend Kwame and my daughter enter a souvenir store, and I decide to walk around a bit. I remember my grandfather telling me a story about losing his small white dog in his hometown when he was about ten years old. Although it must have been terrible for him I like the story because he brought with it conspiracies and suspense. He told me that he had been searching for his dog until eight o'clock at night. The last time grandfather told the story he accused his brother of chasing the dog too long. The most frustrating aspect for my grand-father during the search was that, apart from the fact he did not find his dog, all the people he saw in the village seemed to be licensed to identify him as 'the son of William' rather than his real name Jan. Moreover they also permitted themselves to add moral comments such as 'you should take better care of the dog' or 'dogs are only for responsible owners'.

In the Zuiderzee museum I have the chance to play with the *affects* of this story and re-experience the loss of the dog.[5] This concept of affect is interesting since it is relational and rather re-producible. Felicity Colman produced a convincing definition:

> Affect is the change or variation that occurs when bodies collide, or come into contact. As a body, affect is a knowable product of an encounter, specific in its ethical and lived dimensions and yet it is also as indefinite as the experience of a sunset, transformation or a ghost.[6]

See also the chapter by Soile Veijola in this book for other instances of affect. The museum recreates the aesthetics of the time period in which my grandfather lived as a small boy. The Dutch government started building the dike which disconnected the Zuiderzee from the sea in the 1920 and started preserving the houses from this period and periods before. I can pretend being my grandfather and losing my dog. I want to know what it feels like to lose a dog and experience provincial morality. At the same time I also ask myself: 'Why reconnect with this experience of losing a dog? What is so appealing about this experience that I want to re-live it? Alexander, do you really after 95 years expect to find the dog back in a museum setting?' I think I want to know how it feels to grow up being organized in a rather fixed society and explore its affects.

Here in the museum I can 'play' with these affects. I walk toward a fishing village imagining being a 10-year-old. The weather is grey and it is raining lightly. The village represents the Dutch fishing vil-lage of Urk in 1910 and looks tidy. Volunteers inhabit the village and perform daily life during opening hours. Since it is almost lunchtime they are engaged in the performance of preparing lunch. A group of 20 other visitors are watching the performance. Now I can experi-ence how grandfather felt. I see volunteers dressed in traditional black dresses engaging in their role playing on the top of a dike in small dike houses. I also see visitors making a video about the scene. Suddenly I see a white furry thing behind the black dresses. I hear a mobile phone ringing and see that the white furry thing is a dog, a white dog. The mobile still rings, and the dog is obscured behind someone's black dress. How did grandfather feel in this village? The mobile still rings; someone should turn it off. Turn the mobile off. The tone is familiar; it is a Samsung.

The dog is gone now, but the phone is there, and grandfather is there, irritated since everyone knows so much about him. All is determined already. Stop that phone. Grandpa, I think I found your dog. It is right there behind the dresses. I will help you grandpa. The phone still rings. Pushing away some wet people holding cameras, I receive an elbow in my side and some angry faces. To my relief, the phone stops ringing. I have to get closer to the dog. I can finish grandpa's work and find the dog, which was chased by his brother. The phone rings again. I run forward toward the ladies in the black dresses, disturbing their performance 'preparing lunch'. I ask one of the women about the dog, and she replies that the lunch needs to be ready at 12.30 because the children are coming home from school. She seems to be very angry at me for entering her performance. The phone rings again. Grandpa has lost his dog. Everyone knows grandpa.

I tumble, slip and fall off the dike. Where is the dog? How do the children get their lunch? Pick up the phone! I feel myself rolling down the hillside of the dike. I land in wet grassland. I lie there and look up to the grey sky and the phone rings; it not only rings but also vibrates on my chest. I see a large crowd including the volunteer actors looking down on me; their angry-looking faces hold a faint feeling of satisfaction. I am cold and can feel my t-shirt getting wet. At the same time I'm getting a bit of rest and my heart rate slows down. I realize that I have experienced 'a hospitality space in a clinical condition'.[7] I had the desire to connect with the museum space in a different way than the curator and visitors had intended. However, the configuration as developed by the museum curator does not allow for unintended interactions. In Deleuzian terms, the hospitality space does not allow for temporary de-territorializations.[8] The configuration of visitors, volunteers and infrastructure does not allow disintegrations to happen. This part of the Zuiderzee museum is overly coded, too stratified. Referring to the words of Friedrich Nietzsche, this space is in a clinical condition, an exhausted and degenerating mode of existence.[9] Ironically, a disintegration did happen when I fell off the dike. In short, my desire to try out affects failed. Now my only desire is to warm up. I refuse to look at my phone to see who had called me.

A healthy space

I recognize my friend Kwame and my daughter in the crowd. Slowly I stand up and climb back up onto the dike, ignoring the crowd. I walk toward my friend Kwame and my daughter. The rain is pouring through the small streets in the museum. My feet are wet, my back hurts, my hands are cold and my daughter complains about being bored, hungry and frozen. I can understand the hunger and still see faint traces of peanut butter around the mouth of my friend Kwame. Suddenly we see a sign that reads: 'Welcome'. We hesitantly accept the invitation and enter one of the museum houses. A woman in her mid-fifties welcomes us. She introduces herself as Gusta. She points to the table and invites Kwame and me to sit down. She then invites my daughter to play with the old wooden Dutch toys that make up part of the exhibit. Gusta is a volunteer who 'inhabits' the museum houses. My friend Kwame and I put our wet coats on the peg. Gusta smiles and offers us apple pie. We all three smile back and happily accept the offer. She cuts the pie and pours coffee for me and my friend Kwame and juice for my daughter. 'Are you enjoying the day?' she asks.

I answer that I am trying to experience how life was 80 years ago, but that I'm cold and happy to be inside in the warm living room with a coffee and pie. I also tell her that I'm glad my daughter enjoys playing with the old toys. I see that my daughter has taken two large pieces of apple pie. Then Gusta asks whether I have found what I was looking for. I tell her that it is nice to have a place to rest. Gusta seems happy with the answer and changes the direction of the knife to give me some extra apple pie. I compare the piece of pie with my friend Kwame's. It pleases me enormously that my piece is larger than his. Sometimes I need confirmation that I still count in this world and with the tip of my finger I point to my larger piece of pie. What a glorious feeling.

This glorious feeling does not last long. My friend Kwame tells Gusta that I had entered a troublesome condition and ended up in the wet grass under the dike. He says I had experienced a non-responsive space of hospitality. I knew theories about spaces of hospitality, but I had never heard of a non-responsive space of hospitality. I ask him what he means. My friend indicates that these spaces are sick spaces that can be categorized as being in a clinical state. Kwame continues that this clinical state is a condition within a vitalist ontology. Life itself is the starting of this ontology rather than structures and institutions. Kwame refers to John Pløger:

> vitalism is potentials of becoming existences through a will to something intensively stimulated in social configurations, relations, encounters, interactions, and situations.[10]

In sick spaces, the potentials of becoming are limited, and consequently sick spaces are spaces which reduce each body's power to act and its potential to go on forming new relations,[11] while healthy spaces, by contrast, are defined by Peta Malins as those that 'increase a body's power to form creative, productive relations and which increase its capacity for life'.[12] This understanding of healthy spaces must not be confused with a normative understanding that can be seen as part of the social hygiene of modernizing and modern societies. For Deleuze and Guattari, healthy spaces allow bodies to go on connecting with other bodies, thus creating new flows of desire and undertaking new becomings.[13] So what has happened with me is that I have experienced a decrease in my body's power to form creative and productive relations.

I tell Gusta that I'm very grateful that she offers me the opportunity to experience a responsive space of hospitality whereby the host is not pre-programmed and is enhancing the capacities of life. I already feel better; however, I'm still a bit pale, tired and not satisfied yet. The idea of a responsive space of hospitality does not satisfy me yet since it does not give any directions as to what I can do to influence the space itself. I take a sip of coffee and challenge my friend Kwame for some theoretical insights into the concept of *spaces of hospitality and its potentialities*. My friend Kwame takes the stage and he quotes a number of academics who also thought about opening up spaces of hospitality.

The first would be Orvar Löfgren who refers to vacationing in the following terms:

> A cultural laboratory where people have been able to experiment with new aspects of their identities, their social relations, or their interaction with nature and also to use the important cultural skills of daydreaming and mind-travelling.[14]

The cultural laboratory does not have a pre-given outcome, but as Löfgren states,

> Reciprocal knowledge between tourists and locals might be created, stereotyped categories might be reinforced, or challenged, and social networks might be unmade, remade, inverted or transformed.[15]

My reaction to this is that I like the room for experimentations, but that it is fairly utopian and that the manual to reach this space is missing.

My friend Kwame agrees and indicates that a second theorist, Mustafa Dikeç, might be of help. Dikeç questions conceiving of the relation between host and guest as absolute.[16] According to Dikeç, being hospitable or extending the notion of hospitality does not imply the sovereign power of the host over the guest, but the recognition that we play shifting roles in our engagements, both as hosts and guests. He further argues that hospitality is not about rules of hospitality being implied or predefined through power relations between hosts and guests, but about recognition that we are hosts and guests at the same time in multiple and shifting ways. What I like about this theory are the mutations between host and guest. In this respect, the roles of the host and the guest are not to be preconceived but are mutually constitutive of each other and are relational in nature.[17]

Third, my friend Kwame introduces Heidrun Friese, who indicates that from a historic perspective the concept of hospitality 'originally referred to the identity of the master of the house (i.e., the family, the household) and that this hospitality was bound to a reciprocal exchange, a mandatory pact and, as such, was part of the legal and institutional framework and a specific *politics of hospitality'*.[18]

Friese acknowledges the problematic meaning of the notion of hospitality, where hospitality refers to both the guest or stranger and the enemy and how this already predefines the guest as *being-foreign* and *hostile*.[19] Predefining the stranger in the space of hospitality already closes down the space. Friese asks for opening up spaces of hospitality. I completely agree with Friese, but how can we practically open up spaces of hospitality? My friend Kwame sees that this idea of hospitality as openness/laboratory/not defined is bringing colour back to my cheeks, but I am still visibly unsettled. So he turns to the work of Deleuze. Although Deleuze does not directly write about spaces of hospitality but he does write about *antidotes*.[20] The antidote to this clinical state is a trip to become other, to go beyond. The 'old' self becomes merely a memory, which is revealed when you ask yourself: 'What do you remember when you go beyond?' I admit this sounds rather confusing and perhaps weird. What does Deleuze mean when he writes *go beyond*? And, moreover, beyond what?

My friend Kwame continues with the idea that to become other is to experience the intensities and the affects beyond humans' own conditioning and stratifications,[21] to reconsider the own self and the structural engagements. Becoming other means to become everything one can be and reconstruct an own inner organization. These social experimentations require spaces in which trips are possible and spaces of hospitality should facilitate these becomings. But I still wonder where I can find my cultural laboratory. I feel very small and pose the question, 'How can someone go beyond the *stratified*, *fixed* and *conventional* hospitality relations in spaces of hospitality, dear friends?' Gusta stands up and says, 'You have two hours to address this quest. We will stay here in our museum house and you need to get yourself together and go beyond yourself in this museum. I wish you all the best'.

The quest

It is still raining outside and I'm getting mentally ready for the quest. It is playtime. To address this quest. I feel that I have to become a stranger among other guests in this museum space, someone who has entered as an outsider. From this position I feel I might be able to address this quest. I remember the concept of the *sociological flâneur*, a somewhat cynical description provided by the criticasters of the methods of early nineteenth-century sociologists such as Georg Simmel and Walter Benjamin.[22] Simmel and Benjamin must also have used such a technique.[23] I remember Benjamin's writings about shopping centres in Paris where he would sit on a bench and observe. Later, Paul Lynch, inspired by Simmel, named his research method *sociological impressionism*. Sociological impressionism is drawing up by writing a snapshot of a hospitality space.

I walk around in the museum and search for a place to become a *sociological flâneur* myself. I need a space where I can become an outsider. I see a free bench in front of the rebuilt pharmacy. This will be my spot, and I walk fast toward the bench before someone else takes it. I now sit on the bench trying to move into another dimension, the dimension of a *sociological flâneur*. However, how does this work? I will first try to become calm and close my eyes and breathe deeply. I feel my heart beating quite fast. I sit for five minutes, until my heart beats at a rather normal rate; then open my eyes. But that's it. My flâneuristic eye has not been opened yet to observe new types of interactions. Symbolically speaking, the curtain remains closed. I see all types of families strolling by engaged in a number of interactions. But probably my senses are too inexperienced to observe unaddressed interactions.

I also think about Shamans, medicine men who are able to move to different worlds and ask advice from the spirits. The spirits surely know an answer to my quest. However, I have never been initiated into these techniques of moving into different life worlds. I do not know the rhythms of the drums nor the ingredients for the medicine to hallucinate. I probably can eat five Mars chocolate bars and just sit and ponder the rather simple theories about experience in space, but these probably will not answer my quest. Nevertheless, I think about Joseph Pine and James Gilmore's *experience economy*, which describes the economic offering of experiences. According to Pine and Gilmore, consumers desire experiences, and more and more businesses are responding by explicitly designing and promoting them. They introduced the four realms of an experience; represented by a circle with the sweet spot right in the middle.[24] The four realms are aesthetic, educational, entertainment and escapist. The idea is that the ideal experience has all realms covered and that the market can provide such an experience.

The guest in this respect becomes subject to neoliberal market forces and is transformed into a consumer. The role of the escapist seems interesting; however Pine and Gilmore's explanation refers to escaping from daily life. In other words, does the activity provide the consumer a chance to flee away from the everyday? Today I must admit that the visit has somehow been instructive in the sense that I learned about delirium and entering a Deleuzian clinical state. Regarding the entertainment realm from Pine and Gilmore, I had my entertainment when Gusta challenged me to this quest. The aesthetics are also pretty much okay. The human scale can be found here, and I see how a Dutch village more or less worked 80 years ago. I have received a clear picture of how all the functions in the village are laid out: the bakery, the fishing man, the banker, all located around the church.

The escape realm is fascinating, and I realize that this escape relates very much to the social psychology theory from Seppo Iso-Ahola, whereby compensation is the keyword. People compensate for their work during their leisure time.[25] Simply said, when someone's job is boring, people seek excitement in their leisure and when the job is exciting, people seek an escape from the excitement. For me, the theory also explains how people are fooled by a neoliberal system, which remains in place by providing the right spaces where people can be entertained by consuming the right type of space depending on their type of job. Which makes them fit for complying with the demands of a neoliberal system.

However, this is not the escape I'm looking for. Iso-Ahola reports about compensation and is part of the Neo Marxist school of thought. My quest, in contrast, is about escaping the pre-produced configuration. But to where? Deleuze and Guattari recognize many existences in space. It stops raining, and I see a bench overlooking the Ijsselmeer. I go and sit down on this bench. I think about my quest. It is very similar to the story of the hobbits who are informed by Gandalf that they have to burn a ring in the fires of Mordor.[26] I feel a bit frustrated; at least these hobbits received some sort of direction. They knew what to take where. The hobbits even had helpers in the form of dwarves, elves and others. I am just left here with a quest, lacking all supernatural aid.

John Ronald Reuel Tolkien, the author of *The Lord of the Rings*, lets the hobbits follow the journey of the hero. This route is described in detail by Joseph Campbell in the book *The Hero with a Thousand Faces*, which presents a model for a hero common across a large number of cultures.[27] In his investigation of hero stories, Campbell discovered a pattern that he named the hero's journey. Through a number of states the hero fulfils his/her quest. In short the hero receives a challenge, leaves home, receives helpers, experiences some adventures and then slays the 'dragon'. The hero then enters a process of transformation and returns home. As a matter of course home will never be the same since the hero has transformed.

A large number of movie makers have used the 'hero's journey', most famously George Lucas's *Star Wars*, in which Luke Skywalker undertakes a quest in a galaxy far, far away.[28] Sitting on the bench, I try to feel what it would be like to be Luke Skywalker and wait for the force. But somehow the force does not appear, and I see myself on a bench overlooking the Ijsselmeer, reduced to a guest.

Experiencing hospity

Three words resonating in my mind are escaping, guest and hero – then I suddenly and unexpectedly see a little boy walking toward my bench. I look around for adults but see no one. The toddler smiles at me and quietly climbs up on the bench. Is this boy sent to me by some sort of Star Wars force as a 'helper'? I smile back at the toddler and look to see whether the boy has elves' ears. Then a heated father appears, does not look at me, picks up the toddler and walks away. The toddler still smiles at me. I see that he has no elves' ears but instead has left his pacifier. For five minutes I stare at the pacifier. I imagine myself going after the parents with the pacifier as a starting point.

It would act as a key to open up a new hospitality space, a starting point for exploration and new experience which can be named *hospity*.[29] This hospity is an experience within spaces of hospitality which is not defined yet; the host guest relationship and interactions are not pre-given. This new space can host different becomings with, on one hand, creative becomings whereby the virtual becomes actual and, on the other hand, rather planned becomings whereby the possible becomes real through a process of realization. The process of moving from a virtual to an actual state is what Deleuze refers to as *divergent actualization*.[30] It is an exploration whereby yet unknown 'elements' become connected. According to Deleuze, who drew on the work of Henri Bergson on creative evolution, the process of 'divergent actualization' is the ability of topological forms to give rise to many different physical instantiations.[31]

The process of realization is where tradition comes in, where etiquette prevails – not the most exciting prospect from the perspective of predictability, since it is planned. The host–guest relationship will be subject to divergent actualization as well and will alter quickly. However, we should not forget that this hospity might be a very temporary space of hospitality, as it is probably connected to a myriad of other spaces which have more planned becomings. Regarding the pacifier, if I bring it back to the father, he will probably politely say 'thank you' and we will never reach hospity. But he can also take another route and ask me to participate together in a traditional Dutch game.

I realize that unexpected findings and explorations involve serendipitous processes. In this respect I like to refer to Pek van Andel's definition of serendipity, which is: 'The art of making an unsought finding'.[32]

I like the reference to art since it does appeal to action and state of mind rather than *just luck*. Serendipities do not just happen by luck but, for their becoming, require a keen eye and a responsive attitude. During the process of serendipity the unexpected findings have to be followed by a process of abduction. This process of abduction can be explained as the process of forming an explanatory hypothesis. Many of today's inventions – such as Viagra, penicillin, x-rays and Velcro – are founded through a serendipitous process. Umberto Eco, author of the book *Serendipities: Language & Lunacy,* explains the habits by which we project the familiar onto the strange to make sense of the world.[33]

As a matter of course, this process would also be applicable for hospitality configurations. A serendipitous process can potentially open up the relationship between host and guest and go beyond. This going beyond should be regarded as overcoming the problematic power relationship between host and guest, which was addressed by Friese in the first part of the chapter. Eureka, I think, I have achieved part of the quest. This going beyond is a process of exploration that potentially de-territorializes fixed configurations like host and guests relationships.

The serendipitous process regarding hospitality space has two parts.[34] First, the unsought finding disrupts space, and second, it creates new space. The unexpected finding can be labeled as an *X thing*.[35] It can be regarded as the start of a Deleuzian-Guattarian *line of flight* which disrupts the configuration, including the relationship between host and guest. Such lines of flight initiate de-territorialization processes.[36] The line of flight is the line of change and metamorphosis, which is not organized in a segmentary sequence. It is one associated with change, which usually moves toward reorganization of forms.[37] It refers to the disintegration of the configuration. It is a process of de-construction whereby the configuration falls apart, which can be temporary in nature (whereby the configuration returns) or a permanent disintegration. This X thing might be very small, but it disrupts nevertheless. However during the second process of exploration which is triggered by the X thing, this X thing initiates the production of new space with a different configuration. This space does not have a predictable becoming since it is new. In a hospity condition, the host–guest relationship alters since both guest and host receive voice and are potentially enabled to develop ideas.

I look up and out over the Ijsselmeer and feel energized. Perhaps the Star Wars force has arrived. With one eye I look at the pacifier, which is now dirty with sand. But how, then, can this work today? I have to return to my daughter, Gusta and my friend Kwame, and present how I have gone beyond the hospitality configuration. Just telling them that I looked for the unexpected and started an explorative process won't work. Somehow I have to provide proof that this works. I have to look for an unsought finding, and then I have to open up a new hospitality configuration.

Then it dawns on me that there is some paradoxical element this whole story. Both host and guest have to search for an unsought finding. How can someone search for something that, in principle, is not sought after? Somehow, I believe that one always has to look for paradoxes and controversies because they operate on the edge of chaos, on the edge whereby ordering principles fail and chaos is lurking, the area of the unresolved and traditional ordering processes fail to make sense. Overcoming chaos and solving paradoxes can be regarded as the task of a hero. I'm in search of a new hero in spaces of hospitality. This hero will search for the unsought finding; this finding can serve as a key to open up the host guest configuration. This finding, for example, can be a person during a vacation in Greece who invites you for a weeding. I feel energized and walk from the bench overlooking the Ijsselmeer toward the 'village' for a process of de-territorialization. This place will never be the same. The *untidy guest* is born. The force is with me, and I will find the unexpected and, through this finding, move to hospity. A new action hero is born next to Superman, Spiderman and Mega Mindy.[38] The untidy guest will 'save' both host and guest from the fixed position and performances and clear up clinical states. I wonder whether I need a cape, just like Superman but instead with the letter 'H'. For hospity.

I see an elderly man, around 70 years old, who looks very uptight and frustrated. He probably suffers from a protestant work ethic, even at retirement age. I tell him that he should not worry too much and that the untidy guest will solve all the problems once the unexpected finding is found. At the same I realize that it might be a bit optimistic. To undo 70 years of conditioning with one unsought finding and hospity is quite hard. But perhaps it might have enough force to blow away all the guilt and frustration which has accumulated over the decades. But the good thing about an untidy guest is that he/she does not have to think about the details in an early state. The Holy Grail is to find the unsought.

The untidy guest smiles at fathers and mothers who are struggling with the process of disciplining their children and winks toward the 14-year-old girl who is seemingly annoyed at how her father behaves. Shall this girl be the X thing, the line of flight which opens up new routes? The girl stares back with an angry face. I'm sure she does not recognize me as the superhero: The untidy guest. When I look at myself with a wet jacket and my hair damp around my face I can only agree with her. I should have had the cape. A new hero is born ready to engage in the hero's journey. After 15 minutes of active searching for the unknown I become tired. The hospitality space does not show any cracks where I can find the unknown. I walk past Gusta's home and see my friend Kwame still sitting there drinking his coffee and my daughter playing with the old fashioned toys. No time to watch any longer. The unknown needs to be found. Since this space is in a clinical condition. The guest has become empowered.

Becoming Tijl Uilenspiegel

Then I see a sign announcing an audition for role players in the museum. These are temporary roles for a couple days a year. I see there is one role available as Tijl Uilenspiegel, an ancient Flemish/German street artist, a myth, a Roman figure. Tijl Uilenspiegel is a very controversial figure who performs in villages and tricks people. The museum wants to have a street artist perform in front of the town hall to entertain the visitors. Since I like Tijl Uilenspiegel very much, I become enthusiastic. I almost forget my quest. Perhaps the untidy guest has already found his unsought finding and is ready for hospity. But during the auditions, I fail due to my poor performance on the rope. I was not the only one who was rejected. Three others were also rejected, and to overcome our disappointment we all decided to have a drink together. I say, 'I know a good place where the coffee is warm and the company is pleasant'.

The four of us go to Gusta's place. Gusta seems surprised that I have brought three other guests. She says, 'I am very curious whether you have achieved your quest to see how someone can go beyond the stratified, fixed and conventional hospitality relations in spaces of hospitality'. My friend Kwame is less friendly and says, 'You and your guests cannot come in until you provide us with an acceptable answer on the quest'. I stand up, push my chest forward, and in a formal baritone voice pronounce: 'Through entering lines of flight and a serendipitous mindset one can experiment with the creative potentials in spaces of hospitality'. I say that to go beyond spaces of hospitality, one has to become an untidy guest. This untidy guest invites a serendipitous mindset, finds an unexpected finding (X thing) which triggers a line of flight and engages into a process of experimentation. By entering the line of flight and the experimental process of finding meaning for the X thing, hospity space is created which has no fixed host and guest role, and consequently these roles alter. Hospity space is rather unpredictable.

To my relief, my friend Kwame, who has spent quite a long time with Gusta, nods and says: 'Oh yes, this hospity space relates to Löfgren's concept of the cultural laboratory'. I add that Betsy Wearing articulates that the creative potential of spaces of hospitality can also be addressed by the concept of *chora* which extends the 'I/Me' duality. (See also the chapter by Jennie Germann Molz in this book for the concept of *chora*.) Gusta says that she understands the principle and that unexpected things always make her vacations much more enjoyable. She tells us about a holiday experience where she and her family were unexpectedly invited to a Greek wedding party. She still remembers sitting under the Greek sun and sharing a bottle of wine. Richard, the character from the movie *The Beach*, also harbours a desire to experience something that is not already predetermined, as Olli Pyyhtinen describes in his chapter in this book. And I remember a similar example from the book *Bobos In Paradise: The New Upper Class and How They Got There* by David Brooks in which a man who is bored with being a tourist in Tuscany decides to pretend fainting and hopes that the configuration of being a tourist in Tuscany can be suspended for a while to become a 'someone' in Tuscany.[39] The door of a house opens and a worried person takes the man in. The man in David Brooks' book bridges the gap between being a tourist and being someone. He searches for the creative potential in spaces of hospitality by entering a disruptive process. He finds the X thing in the form of falling down and temporarily escapes the stratified, fixed and conventional hospitality configurations in spaces of hospitality. We stay a while, and I hear a security officer knocking on the door; I hear him shouting that the museum is closed already; I see Gusta looking around the room, she unexpectedly has a great number of people around the table, and I try to remember what has happened during the day. I close my eyes and see, vaguely, peanut butter sand-wiches, mobile phones, lost dogs, sandy pacifiers and a number of Tijl Uilenspiegels marching through the inside of my eye.

Conclusion

Deleuze and Guattari encourage scholars to create their own con-cepts to grasp contemporary becomings. Moreover, they go as far as to encourage authors to create their own interpretation and adapta-tion of Deleuzian-Guattarian thinking *through* Deleuzian-Guattarian thinking. I like this idea very much.[40] Their concepts are meant to be changed and personalized in order to address contemporary life. New concepts need to be developed since contemporary life cannot be completely described with yesterday's concepts. I have stressed that spaces of hospitality need to encourage and facilitate hospity experiences. Spaces of hospitality should facilitate X things which initiate lines of flights so that both the host, as well as the guest, can escape stratified, fixed and conventional hospitality configurations and explore what is beyond. X things and hospity are new concepts that can be used in the discussion about spaces of hospitality. New futures become visible, and can be visited and experienced. Then, spaces of hospitality become cultural laboratories[41] in which people can experiment with alternative futures and identities. My story shows Gusta organizing her space of hospitality within the museum space. I have played with 'hosting' my grandfather's identity, Tijl Uilenspiegel's identity and invited a serendipitous mindset. I have even invited three other wannabe Tijl Uilenspiegels into Gusta's space of hospitality. In Deleuzian terms, host and guests enter a process of de-territorialization of the hospitality space through following a line of flight. The space created during the de-territorialization is called hospity. In time the original space re-territorializes. The untidy guest restlessly searches for the line of flight and plays with its actualization of potentialities.

From a neo-vitalistic Deleuzian perspective, spaces of hospitality are alive; consequently, they can also be evaluated clinically. Whereas the French philosopher Deleuze uses the concepts of the clinical and critical to evaluate the becoming of life forces in literature, I use the concepts of the clinical and the critical to evaluate spaces of hospitality as a 'physician of culture'.[42] For the diagnosis, scholars have to look at phenomena of culture. Drawing on Nietzsche,[43] Deleuze regards phenomena as 'signs or symptoms that reflect a certain state of forces'.[44] The criteria for addressing healthy states are, according to Deleuze, on the one hand, *a bad sickly life*, which is an 'exhausted and degenerating mode of existence, a life that hinders life energies'. The *good or healthy life*, by contrast, is what Deleuze refers to as an 'overflowing and ascending form of existence, a mode of life that is able to transform itself depending on the forces it encounters, always increasing the power to live, always opening up new possibilities of life'.[45] Nietzsche, in this respect, uses the German term *Mehr-Leben*, which translates as more-life.[46]

Spaces of hospitality *in potentia* have great opportunities for connecting the unknown and become spaces of difference which intensify and transmit life rapidly so from a vitality perspective are highly healthy spaces. The concept of an untidy guest becomes a methodology for addressing vitality in spaces of hospitality.

And now, back to our diagnosis. Is the space of hospitality a healthy space? When hosts and guests are able to find the unexpected, engage in a trajectory of exploration and experience hospity, the space of hospitality can be considered *a healthy space of hospitality*. A relevant question in this respect is whether the host and guest can organize temporary space of hospitality within another space of hospitality whereby bodies, ideas and memories become connected. In other words 'host and guest organise their own party', as embodied by Gusta and her guests in the story. The potentiality of finding X things and entering a process of experimentation in spaces of hospitality informs about the clinical condition of the space of hospitality.

Spaces of hospitality such as museums, restaurants and amusements parks have the potential to become spaces that enhance life itself. They can become spaces that allow exploring future potentialities. The notion of healthy spaces of hospitality implies that spaces of hospitality allow us to transcend the host-guest dichotomy, that spaces of hospitality can open up difference and enter host and guest experience hospity. The invitation of a serendipitous and creative mindset already initiates a different kind of hospitality space and is also an actualization of finding ways of working toward a being-with.

Notes

1. Gilles Deleuze and Felix Guattari, *A Thousand Plateaus: Capitalism and Schizophrenia*, translated by B. Massumi, Minneapolis, University of Minnesota Press, 1987, p. 97.
2. Ibid., p. 98.
3. Translated from Dutch to English by author.
4. See also http://www.zuiderzeemuseum.nl. Accessed 10 March 2014.
5. Deleuze and Guattari, *A Thousand Plateaus*, p. 9; among others.
6. Felicity Coleman, 'Affect', in A. Parr (ed.), *The Deleuze Dictionary*, New York, Columbia University Press, 2005, pp. 11–15.
7. The concept of the clinical state is based on the work of Gilles Deleuze, *Essays Critical and Clinical*, translated by M. A. Greco and D. W. Smith, Minneapolis, University of Minnesota Press, 1998, p. xiv.
8. See also the discussion of spaces of hospitality in Alexander Grit and Paul Lynch, 'An Analysis of the Development of Home Exchange Organisations', *Research in Hospitality Management*, 1.1, 2012, pp. 1–7.
9. Gilles Deleuze, *Essays Critical and Clinical*, translated by Michael A. Greco and Daniel W. Smith, Minneapolis, University of Minnesota Press, 1998, p. xiv.

10. John Pløger, *In Search of Urban Vitalis*, Space and culture vol. 9 no. 4, 2006, pp. 386.

11. Concept drawn from the work of Peta Malins, 'Machinic Assemblages: Deleuze, Guattari and an Ethico-Aesthetics of Drug Use', *Janus Head*, 7.1, 2004, pp. 97–98.

12. Ibid., p. 97.

13. Concept drawn from the work of Deleuze and Guattari, *A Thousand Plateaus*.

14. Orvar Löfgren, *On Holiday: A History of Vacationing*, Berkeley, University of California Press, 2002, p 7.

15. Ibid.

16. Dikeç, Mustafa, 'Pera Peras Poros: Longings for Spaces of Hospitality', *Theory, Culture & Society*, 19.1–2, 2002, p. 239.

17. Ibid.

18. Heidrun Friese, 'Spaces of Hospitality', *Journal of Theoretical Humanities*, 9.2, 2004, p. 69.

19. Ibid.

20. The concept of antidotes is extensively described in the introduction of Deleuze, *Essays Critical and Clinical*.

21. The author is indebted to the Deleuze classes in 2013 in Utrecht from contemporary philosopher and feminist theoretician Rosi Braidotti.

22. This is a form of being in the world signifying strolling, idling, often with the idea of wasting time.

23. Describing urban experience in a sociological and psychological language. See also Georg Simmel's essay, 'The Metropolis and Mental Life', in Donald Levine (ed.), *Simmel: On Individuality and Social Forms*, Chicago, Chicago University Press, 1971, p. 324.

24. B. Joseph Pine and James H. Gilmore, *The Experience Economy: Work Is Theatre & Every Business a Stage*, Boston, Harvard Business School Press, 1999.

25. Seppo Iso-Ahola, 'Toward a Social Psychological Theory of Tourism Motivation: A Rejoinder', *Annals of Tourism Research*, 9.2, 1982, pp. 256–262.

26. John Ronald Reuel Tolkien, *Lord of the Rings*, London, Harper Collins, 1954.

27. Joseph Campbell, *The Hero's Journey*, New York, Bollingen, 1949. In this book the path of the hero is described.

28. *Star Wars: Episode IV – A New Hope*, screenplay and director George Lucas, Lucasfilm and Twentieth Century Fox, 1977.

29. Alexander Grit and Paul Lynch, 'Moving Towards Hospity', 2013, unpublished paper. I am thankful to Sara Baranzoni, Armand Gouvernante and Maaike de Jong for some inspiring dialogue on this topic at the Deleuze workshop March 2012, Manchester Metropolitan University.

30. Gilles Deleuze, *Difference and Repetition*, translated by P. Patton, New York, Columbia University Press, 1994, p. 168.

31. Gilles Deleuze, *Bergsonism*, translated by H. Tomlinson and B. Habberjam, New York, Zone, 1991. See Alexander Grit and Paul Lynch, 'Hotel Transvaal and Molar Lines as a Tool to Open Up Spaces of Hospitality',

in Irena Ateljevic, Nigel Morgan and Annette Pritchard (eds), *The Critical Turn in Tourism Studies: Creating an Academy of Hope*, London, Routledge, 2011, pp. 208–218.

32. Pek van Andel, 'Anatomy of the Unsought Finding Serendipity: Origin, History, Domains, Traditions, Appearances, Patterns and Programmability', *British Journal for the Philosophy of Science*, 45.2, 1994, p. 632.

33. Umberto Eco, *Serendipities: Language and Lunacy*, New York, Columbia University Press, 1998.

34. The concept of serendipitous hospitality experiences has been described in: Alexander Grit, *The Opening Up of Hospitality Spaces to Difference: Exploring the Nature of Home Exchange Experiences*, unpublished Ph.D. thesis, Department of Management, University of Strathclyde, 2010.

35. Ibid. The term X thing is used by the author in the thesis *The Opening Up of Hospitality Spaces to Difference: Exploring the Nature of Home Exchange Experiences* to address the 'yet unknown'.

36. Ibid., pp. 9–10. 'Lines of flight' is a Deleuzian-Guattarian term that refers to bolts of pent-up energy that break through the cracks in a system of control and shoot off on the diagonal. By the light of their passage, they reveal the open spaces beyond the limits of what exists.

37. Ibid., pp. 9–10.

38. Mega Mindy is a Flemish superhero in a television series for children.

39. David Brooks, *Bobos in Paradise: The New Upper Class and How They Got There*, New York, Simon & Schuster, 2010.

40. See also Senija Causevic and Paul Lynch, 'Phoenix Tourism: Post-Conflict Tourism Role', *Annals of Tourism Research*, 38.3, 2011, pp. 780–800, where the authors introduce the new concept of 'phoenix tourism' to address an experience in space in hospitality.

41. Löfgren, *On Holiday*, p. 7.

42. Deleuze, *Essays Critical and Clinical*, p. xiv.

43. Ibid. The author distances himself from any Italian and German fascist regime which claimed Nietzsche's ideas regarding vitalism. The author in this respect places himself in a neo-vitalistic tradition, which is highly critical toward fascists' interpretations of vitalist ontology.

44. Ibid.

45. Malins, *Machinic Assemblages*, p. 97.

46. Gilles Deleuze, *Nietzsche and Philosophy*, translated by J. Tomlinson, New York, Columbia University Press, 1983.

7
Conclusion: Prepositions and Other Stories

The camping experiment is soon coming to an end. We are packing our tents and kettles, and are ready to return home. One more time we sit down around the campfire and reflect on what has happened. Have we provided well-placed stepping stones through the theoretical and literary landscapes to support ontological reconsiderations in tourism and hospitality research, as we had aspired to do? Are there more spaces of being and spaces of being-with, perhaps? Have we left enough lines of flight for the tourists of the future?

We have dwelled in the company of 'the untidy guest' throughout our book in order to disrupt established frames in hospitality as well as in tourism theory and practices. However, far from presenting a unified concept that would bring all things into unison, the notion of the untidy guest been created anew in each of the chapters. All of them varied or mutated the concept in relation to a specific theme and the pairs of concepts they introduced – camping and clearing, paradise and parasite, silence and community, unlearning and hospitality, messing around and serendipity. Thus, there is no *one* meaning which the notion would denote, nor an exhaustive definition of it. The untidy guest appears as one and multiple at the same time; while being a subject of common being, it came up as slightly different in each of the chapters.

The same applies to the idea of disruptive tourism. While each chapter opened up a social world of its own related to mobile and hospitable living and dwelling, the chapters shared the conceptual premise of 'disruption' as a way to rethink the rules and ways of knowing, experiencing and doing hospitality. Instead of defining tourism, hospitality, sustainability or ethics through essentialist categorizations, we have transferred the focus of tourism theorizing towards alternative ontologies and epistemologies of tourism. By this we mean that ethics cannot be pre-designed or assumed without taking notice of 'the social', without openness towards others. Namely, even the most sophisticated 'codes of ethics' for tourism do not necessarily bear significance in real life.[1] Embracing both untidy guesting and untidy hosting advocates a readiness to embrace openness and withness in relation to 'the other'.[2] Our intervening concepts – camping, parasite, silence, unlearning and serendipities – are constituents of travelling; they are not items to be packed and marketed for pre-decided and pre-designed tourist destinations. The approaches of disruptiveness call for acknowledging and cherishing the *im*possibility of rendering the good life or its production processes into a marketable commodity.

Throughout the book, we have argued for new or, rather, alternative ontologies of tourism and hospitality in theory and practice. The alternative ontologies are about relations; they reconfigure the relations between host and guest, and the forms of tourist encounters. Thereby *prepositions* have presented themselves to us as workable tools to examine such relations, with each preposition giving an expression to a different relationship. In an interview, Michel Serres once suggested that traditionally philosophy has spoken in substantives or verbs. In contrast to this, he says, his own philosophy operates by way of prepositions, such as between, with, across and beside.[3] For Serres, prepositions are, literally, *pre*-positions; they indicate relations that precede any fixed positions.[4]

In our book, we have proceeded with a similar idea and developed alternative ontologies for tourism and hospitality by remembering the power of prepositions. The chief preposition that we have worked with is, undoubtedly, *with*. The chapters render questions about being into questions about being-with. As Soile Veijola noted in her chapter on silent communities, we take up the idea formulated by both Georg Simmel and Jean-Luc Nancy that being is always given and put into play as being-with. I am necessarily along with others; there are no occasions of being where being would not be constituted by withness. Even being alone, as both Simmel and Martin Heidegger suggest, is a form being-with, albeit a deficient form.[5]

Here, rather than giving an expression to a unique relationship, the preposition 'with' is one that brings all relations together. Insofar as 'being' always takes the form of 'being-with', being-with does not have to do only with harmonious sharing, but it can also be the form of 'being-against' in competition, conflict, war or struggle. Simmel was perceptive of the fact that also those forms of relations which oppose individuals or groups to one another equally bind them together.[6] 'It is precisely being-with that is played out also in the forms of being-against'.[7]

Therefore, while we commence in the book from the idea that being is always being-with, we do not assume that relating to the other in encounters would happen easily, just like that. On the contrary, making connections, especially *good* ones, and making them hold together is a difficult, laborious and uncertain task. One easily fails in it. Hence being-with requires effort, prudence and care. It is precisely because of the fact that connecting is arduous and costly that the move to bring ethics in, and 'explore ontological being as ethical togetherness', as Veijola coins it in her chapter, is so important. While designating the ontological precondition of being, the preposition 'with' therefore also indicates a specific kind of relation. It has to do with ethical togetherness, that is, with welcoming the untidy guests and finding ways to be together peacefully with them. For Gilles Deleuze, the ethical is directly linked to the potentiality to form new connections in the future. These connections are able to transmit life itself in an intensive manner. The host–guest relationship underlying the ethical togetherness allows thus for future explorations, with unanticipated outcomes, with others – unless, of course, the host–guest relations are thoroughly managed and governed into 'a behavior' of a tourist, a host or a tourism employee.

To put it in concrete terms, a set of other prepositions are worked with in the chapters, as indicated in their titles: Camping *in* Clearing, Paradise *with/out* Parasites, *Towards* Silent Communities, Unlearning *through* Hospitalities and Messing *around with* Serendipities. In our book, we are in, with/out, towards, through and around with others. Hereby we have directions, durations, horizons and relations in our being-with.

In the chapter on camping, Jennie Germann Molz introduced Heidegger's ontological notion of clearing as a form of being and becoming. The phrase 'camping *in* clearing' echoes the relationship between dwelling and place that we find in Heidegger, namely that we dwell *in* the place. But clearing disrupts and it also invites disruption. It is a way of being in the world that is perhaps too tidy and too solitary for the playful possibilities of being-together. Camping *in* clearing thus connotes a kind of snug containment that belies the excesses, leakages and violences of the clearing. It is for this reason that Germann Molz played around with other, somewhat messier, concepts in order to fray the edges of the clearing. Through the concepts of *chora* and unfinishedness, she argued, we are able to inhabit new ways of being together with difference. From this perspective, 'being in' might be rethought as 'being with', 'being along'[8] or 'being in-between' – prepositional relations that leave the door open, so to speak. After all, it is in between ourselves and others that we find the generative capacity of *chora*. As the story of camping unfolded in the chapter, it became clear that camping refuses bounded completeness and instead flourishes in unfinishedness.

In the chapter on the paradise and parasites of the movie *The Beach*, Olli Pyyhtinen examined how community is only made possible by the exclusion of the parasite. The parasite is the one who or that which inter-feres, inter-venes or inter-rupts, and is thus signalled by the preposition 'between'. It is a third standing in-between, while also at the same time being 'beside' or next to the host. However, the chapter also argued that the parasite not merely interferes relations but it is also their precondition. The parasite is not external to a system, merely a transitory, marginal nuisance, but part of the system itself. It is at once necessary for the system and an obstacle for its proper functioning. This duality was also expressed by the expression *with/out* in the title: there is no paradise with parasites, and there is no paradise without parasites. The slash in the middle, as an in-between that is parasitic in its own right, effectuates a cut and acts as a dividing line separating the ordered inside from the chaotic and dis-ordered outside. It marks a border or threshold preventing one from passing from one side to the other, neither in one direction or the other, without being interrogated and without having the right credentials. However, towards the end of his chapter Pyyhtinen also stressed the pressing need to interrupt exclusion, which is always an act of violence, and find ways of co-existing and living with the parasite. Such co-existence designates a space of novelty and presents a precondition of ethics.

In the chapter on silent communities, Soile Veijola gave as much rope as possible to the being-with-others while being-with-oneself. For the experiment, she used the notion of the silent community, arguing that by way of its constitutive *towardness* and *withwardness*, the potentialities of both being-oneself and being-together find expression and meaning equally. Silence is thereby a state of being-with that, in its very precariousness, illuminates the ethical aspects of the personal, social and the communal; it is never settled or solved for good, nor can it be defined or measured technically. Neither can it be decided or imposed upon people without becoming a silenced community which would, of course, be a coercive or violent one. Yet it can feed creativity and the unexpected as an obstruction to one's habitual living with others. Silence and the silent community are untidy concepts that can at best be approached and lived through by embracing them as incessantly transforming modes of ethical engagement in being-with.

In the chapter on coffee trails in the Nicaraguan highlands, Emily Höckert discussed the ways in which subjectivities could be transformed and redefined with the help of the pivotal thought of hospitality. During the field trip, the preposition *through* allowed the travellers to disrupt the taken-for-granted limits between self and other. Höckert's narrative unloaded the backpacks of researchers and development workers in order to make more space for intersubjective ontologies. She turned her topic of learning and teaching in development encounters into a play through which one could both learn and unlearn. Borrowing the notion of hospitality from the basic toolbox of tourism studies, Höckert drew special attention to Derrida's and Levinas' ideas of the prepositions between *welcome* and *the other*. These authors encourage one to envision more equal and ethical encounters where the 'welcome *of* the other' always precedes 'welcome *to* the other', meaning that it is always 'the other' who can say the first 'yes'. Hence, hospitality and welcome interrupt oneself and the tradition of autonomous, egological subject through giving priority and infinite space to messiness and reciprocity.

In the narrative of visiting the Zuiderzee museum in the Netherlands, Alexander Grit played *around with* Deleuze's concepts of the clinical and the critical to address the health of the spaces of hospitality.[9] Spaces of hospitality are complex spaces in the sense that they host an array of different relationships. Usually the nature of these relationships prevents and discourages processes of messing, playing and dancing *around with*. But serendipitous processes can open up space and enable an experimentation with different futures.[10] The title 'messing around with serendipities' radiates the idea that there are potentialities in spaces of hospitality that must be brought into being. Healthy spaces of hospitality should invite both hosts and guests to dance *around with*, to play *around with*, to experiment *around with* and to mess *around with* the other. The preposition *around* suggests a middle – something around which we dance, play, experiment and mess – but that middle is not necessarily fixed. On the contrary, it entails the ever-changing potentialities of spaces and relations of hospitality. With the concept of hospity, Grit argued that in optimally healthy spaces of hospitality, host and guest roles can shift; the other can be invited; and future potentialities can be explored.[11]

To summarize, we understand being-with also in a methodological sense, as a matter of exposure to the messiness, openness and unpredictability of tourism and hospitality rather than as an effort to clean it up or stamp it flat. This said, a book like ours cannot be made – nor read – without replacing 'the logics of one single rule'[12] with the creativity fed by both freedom and obstructions.

We have one more story to share before we extinguish the camp-fire. Do you have a minute?

You remember Charles Baudelaire, the poet, whose poem 'Anywhere Out of the World' (*N'importe où hors du monde*) from the famous *Le Spleen de Paris* published in 1862 opened that classic text of tourism studies, *The Tourist* by Dean MacCannell?[13] Baudelaire had found out about a place far up in the north called 'Torneo', in one of those countries 'analogous to Death' towards which his 'poor chilled soul' could flee. As it happens, the city of Tornio lies by The Tornionjoki river bordering Sweden and Finland. In the currents of the very same river, visual artist Markku Akseli Heikkilä has a little while ago experimented with the encounter of cream and water. Wanting to bring cream to pay a symbolic visit to the small, sandy islands that used to host cows and calves in the summers of his childhood, he witnessed to his surprise the two substances temper each other without dissolving into one another, without becoming one. Cream and water in the river turned into peaceful and beautiful co-existence that looks like the softest smoke, which you can now see on the cover of our book – if you have the heart to close it.

Notes

1. David Fennell, *Tourism Ethics*, Clevedon, Channel View Publications, 2006, p. 296; see also Mick Smith, 'Ethical Perspectives: Exploring the Ethical Landscape of Tourism', in Tazim Jamal and Mike Robinson (eds.), *The SAGE Handbook of Tourism Studies*, London, SAGE, 2009, pp. 625–629.
2. Rauna Kuokkanen, *Reshaping the University. Responsibilities, Indigenous Epistemes, and the Logic of the Gift*, Vancouver, UBC Press, 2007, p. 139.
3. Michel Serres with Bruno Latour, *Conversations on Science, Culture, and Time*, translated by R. Lapidus, Ann Arbor, University of Michigan Press, 1995, p. 101.
4. Ibid., p. 105.
5. Georg Simmel, *Soziologie*, in Georg Simmel Gesamtausgabe, Band 11, Frankfurt am Main, Suhrkamp, 1908/1992, p. 96; Martin Heidegger, *Sein und Zeit*, Max Niemeyer, Tübingen, 1927/1972, §26 p. 118.
6. Simmel, *Soziologie*, e.g. pp. 284–288.
7. Olli Pyyhtinen, 'Being-with: On Georg Simmel's Sociology of Associations', *Theory, Culture & Society*, 26.5, 2009, p. 116.
8. Nuccio Mazzullo and Tim Ingold, 'Being Along: Place, Time and Movement Among Sámi People', in Juergen Ole Baerenholdt and Brynhild Granås (eds), *Mobility and Place: Enacting Northern European Peripheries*, Aldershot, Ashgate, 2008, pp. 27–38.
9. Gilles Deleuze, *Essays Critical and Clinical*, translated by D. W. Smith and M. A. Greco, London, Verso, 1993/1998.
10. Alexander Grit, *The Opening Up of Hospitality Spaces to Difference: Exploring the Nature of Home Exchange Experiences*, unpublished PhD thesis, University of Strathclyde, 2010.
11. Alexander Grit and Paul Lynch, 'Moving Towards Hospity', unpublished paper, 2013.
12. Michael Hardt and Antonio Negri, *Empire*, Cambridge, MA, Harvard University Press, 2000, p. xii.
13. Dean MacCannell, *The Tourist: A New Theory of the Leisure Class*, Berkeley, University of California Press, 1976.

Bibliography

Adler, J. (2002) 'The Holy Man as Traveler and Travel Attraction: Early Christian Asceticism and the Moral Problematic of Modernity', in W. H. Swatos, Jr. and T. Luigi (eds), *From Medieval Pilgrimage to Religious Tourism: The Social and Cultural Economics of Piety*, Westport, CT, and London: Praeger, pp. 25–50.

Agamben, G. (1990/2007) *The Coming Community*, translated by M. Hardt. Minneapolis, MN: University of Minnesota Press.

Agamben, G. (1998) *Homo Sacer: Sovereign Power and Bare Life*, translated by D. Heller-Roazen, Stanford, CA: Stanford University Press.

Agamben, G. (2005) *State of Exception*, Chicago, IL: Chicago University Press.

Ahmed, S. (2000) *Strange Encounters: Embodied Others in Postcoloniality*, London: Routledge.

Aitchinson, C. (2001) 'Theorizing Other Discourses of Tourism, Gender and Culture. Can the Subaltern Speak in (Tourism)?', *Tourism Studies*, 1.2, pp. 133–147.

Amin, A. and Thrift, N. (2002) *Cities: Reimagining the Urban*, Cambridge: Polity Press.

Andel, P. (1994) 'Anatomy of the Unsought Finding Serendipity: Origin, History, Domains, Traditions, Appearances, Patterns and Programmability', *British Journal for the Philosophy of Science*, 45.2, pp. 631–648.

Ankor, J. and Wearing, S. (2012) 'Gaze, Encounter and Philosophies of Otherness', in O. Moufakkir and Y. Reisinger (eds), *The Host Gaze in Global Tourism*, Oxfordshire: CABI, pp. 179–191.

Appiah, K. A. (2006) *Cosmopolitanism: Ethics in a World of Strangers*, New York: W.W. Norton Company.

Augé, M. (1995) *Non-places: An Introduction to Supermodernity*, London: Verso.

Bachelard, G. (1958/1994) *The Poetics of Space*, translated by M. Jolas, Boston, MA: Beacon Press.

Backman, J. (2013) 'Olemisen ainutkertaisuudesta ainutkertaisuuden politiik-kaan: Parmenides, Heidegger, Nancy' [From the Singularity of Being to the Politics of Singularity: Parmenides, Heidegger, Nancy], *Tiede & Edistys*, 2, pp. 108–124.

Baker, G. (2011) *Politicising Ethics in International Relations: Cosmopolitanism as Hospitality*, London: Routledge.

Baker, G. (ed.) (2013) *Hospitality and World Politics*, New York: Palgrave Macmillan.

Bell, D. (2007) 'Moments of Hospitality', in J. Germann Molz and S. Gibson (eds), *Mobilizing Hospitality: The Ethics of Social Relations in a Mobile World*, Aldershot: Ashgate, pp. 29–45.

Bell, D. (2009) 'Tourism and Hospitality', in T. Jamal and M. Robinson (eds), *The SAGE Handbook of Tourism Studies*, London: Sage, pp. 19–34.

Benveniste, É. (1973) *Indo-European Language and Society*, translated by E. Palmer, London: Faber and Faber Limited.

Benvenuto, S. (2000) 'Fashion: Georg Simmel', *Journal of Artificial Societies and Social Simulation*, 3.2, http://jasss.soc.surrey.ac.uk/3/2/forum/2.html.

Bergeå, J. (2002) *Class Travellers in Chicken Buses, Backpacking, Economic Inequality and Cultural Identity in Guatemala*. Minor Field Study, Gothenburg University, Peace and Development Research Institute. http://www.resurs. folkbildning.net/PageFiles/10619/class%20travellers.PDF.

Berking, H. (1999) *Sociology of Giving*, translated by P. Camiller, London: Sage.

Birkeland, I. (2005) *Making Place, Making Self: Travel, Subjectivity and Sexual Difference*, Aldershot: Ashgate.

Birtchnell, T. and Büscher, M. (2011) 'Stranded: An Eruption of Disruption', *Mobilities*, 6.1, pp. 1–9.

Bissell, D. (2013) 'Pointless Mobilities: Rethinking Proximities through the Loops of Neighbourhood', *Mobilities*, 8.3, pp. 349–367.

Bissell, D. and Fuller, G. (eds) (2011) *Stillness in a Mobile World*, London and New York: Routledge.

Boellstorff, T. (2011) 'Placing the Virtual Body: Avatar, Chora, Cypherg', in F. E. Mascia-Lees (ed.), *A Companion to the Anthropology of the Body and Embodiment*, Malden, MA: Blackwell.

Boorstin, D. (1964) *The Image: A Guide to Pseudo-Events in America*, New York: Harper.

Bourdain, A. (2012) 'Japan: Cook it Raw', *No Reservations* (television program), Travel Channel, 7 May 2012, Season 8, Episode 5.

Brah, A. (1996) *Cartographies of Diaspora: Contesting Identities*, London: Routledge.

Briggs, J. and Sharp, J. (2004) 'Indigenous Knowledges and Development: A Postcolonial Caution', *Third World Quarterly*, 2.4, pp. 661–676.

Buber, M. (1923) *Ich und Du*. Leipzig: Im Insel.

Campbell, J. (1949) *The Hero with a Thousand Faces*, New York: Bollingen.

'Camping', *Parks and Recreation* (television program), NBC, 24 March 2011, Season 3, Episode 8.

Cañada, E. (2010) 'Perspectivas del Turismo Comunitario: Cómo Mantener Vivas las Comunidades Rurales', http://blog.pucp.edu.pe/item/93900/ mundo-perspectivas-del-turismo-comunitario-como-mantener-vivas-las- comunidades-rurales.

Causevic, S. and Lynch, P. (2011) 'Phoenix Tourism: Post-Conflict Tourism Role', *Annals of Tourism Research*, 38.3, pp. 780–800.

Chalip, L. and Costa, C. A. (2012) 'Clashing Worldviews: Sources of Disappoint-ment in Rural Hospitality and Tourism Development', *Hospitality & Society*, 2.1, pp. 25–47.

Cocker, E. (2011) 'Performing Stillness: Community in Waiting', in D. Bissell and G. Fuller (eds), *Stillness in a Mobile World*, London and New York: Routledge, pp. 87–106.

Cohen, E. (1979) 'The Impact of Tourism on the Hill Tribes of Northern Thailand', *Internationales Asienforum*, 10.1/2, pp. 5–38.

Cohen, G. A. (2009) *Why Not Socialism?*, Princeton and Oxford: Princeton University Press.

Coleman, F. (2005) 'Affect', in A. Parr (ed.), *The Deleuze Dictionary*, New York: Columbia University Press, pp. 11–14.

Conradson, D. (2011) 'The Orchestration of Feeling: Stillness, Spirituality and Places of Retreat', in D. Bissell and G. Fuller (eds), *Stillness in a Mobile World*, London and New York: Routledge, pp. 71–86.

Cresswell, T. (1996) *In Place Out of Place: Geography, Ideology, and Transgression*, Minneapolis, MN: University of Minnesota Press.

Critchley, S. (1999) *The Ethics of Deconstruction. Derrida and Levinas*, Edinburgh: Edinburgh University Press.

Csikszentmihalyi, M. (1975) *Flow: The Psychology of Optimal Experience*, New York: Harper Perennial.

Daly, T. (2010) 'The Intersubjective Ethic of Julieta Paredes' Poetic', *Bolivian Studies Journal, Revista de Estudios Bolivianos*, 15–17, pp. 237–263.

Dann, G. (1999) 'Writing out the Tourist in Space and Time', *Annals of Tourism Research*, 26.1, pp. 159–187.

Deleuze, G. (1966) *Bergsonism*, translated by H. Tomlinson and B. Habberjam, New York: Zone.

Deleuze, G. (1978/2008) 'Deleuze/Spinoza, Cours Vincennes – 24/01/1978'. *Les Cours de Gilles Deleuze*, www.webdeleuze.com.

Deleuze, G. (1983) *Nietzsche and Philosophy*, translated by H. Tomlinson, New York: Columbia University Press.

Deleuze, G. (1994) *Difference and Repetition*, translated by P. Patton, New York: Columbia University Press.

Deleuze, G. (1995) *Negotiations. 1972–1990*, translated by M. Joughin, New York: Columbia University Press.

Deleuze, G. (1998) *Essays Critical and Clinical*, translated by D. W. Smith and M. A. Greco, London: Verso.

Deleuze, G. and Guattari, F. (1987) *A Thousand Plateaus: Capitalism and Schizophrenia*, translated by B. Massumi, Minneapolis, MN: University of Minnesota Press.

Deleuze, G. and Guattari, F. (1994) *What Is Philosophy?* translated by H. Tomlinson and G. Burchell, New York: Columbia University Press.

Derrida, J. (1976) *Of Grammatology*, translated by G. Spivak, Baltimore, MD: Johns Hopkins University Press.

Derrida, J. (1999) *Adieu to Emmanuel Levinas*, translated P.-A. Brault and M. Naas, Stanford, CA: Stanford University Press.

Derrida, J. (2000) *Of Hospitality. Anne Dufourmantelle invites Jacques Derrida to Respond*, translated by R. Bowlby, Stanford, CA: Stanford University Press.

Derrida, J. (2005) 'The Principle of Hospitality', *Parallax*, 11.1, pp. 6–9.

Dicks, B. (2004) *Culture on Display: The Production of Contemporary Visitability*, Maidenhead: McGraw Hill.

Diekmann, A. and Hannam, K. (2012) 'Touristic Mobilities in India's Slum Space', *Annals of Tourism Research*, 39.3, pp. 1315–1336.

Dikeç, M. (2002) 'Pera Peras Poros: Longings for Spaces of Hospitality', *Theory, Culture & Society*, 19.1–2, pp. 227–247.

Diken, B. and Laustsen, C. B. (2005) *The Culture of Exception: Sociology Facing the Camp*, London and New York: Routledge.

Diprose, R. (2011) 'Building and Belonging Amid the Plight of Dwelling', *Angelaki: Journal of Theoretical Humanities*, 16.4, pp. 59–72.

Doel, M. (2000) *Un-glunking Geography: Spatial Science Dr. Seuss and Gilles Deleuze*, in M. Crang and N. Thrift (eds), *Thinking Space*, New York: Routledge, pp. 117–135.

Douglas, M. (1966/2003) *Purity and Danger*, London and New York: Routledge.

Drabinski, J. E. (2011) *Levinas and the Postcolonial. Race, Nation, Other*, Edinburgh: Edinburgh University Press.

Dredge, D. and Hales, R. (2012) 'Community Case Study Research', in L. Dwyer, A. Gill and N. Seetaram (eds), *Handbook of Research Methods in Tourism: Quantitative and Qualitative Approaches*, Cheltenham and Northampton, MA: Edward Elgar, pp. 417–437.

Duffy, R. (2002) *A Trip Too Far: Ecotourism, Politics, and Exploitation*, London: Earthscan.

Dussel, E. (2008) *Twenty Theses on Politics*, translated by G. Ciccariello-Maher, Durham: Duke University Press.

Eco, U. (1998) *Serendipities: Language and Lunacy*, New York: Columbia University Press.

Edensor, T. (2013) 'Reconnecting with Darkness: Gloomy Landscapes, Lightless Places', *Social & Cultural Geography*, 14.4, pp. 446–465.

Eriksson Baaz, M. (2005) *The Paternalism of Partnership. A Postcolonial Reading of Identity in Development Aid*, London: Zed Books Ltd.

Escobar, A. (2010) 'Latin America at a Crossroads', *Cultural Studies*, 24.1, pp. 1–65.

Escobar, A. (2012) *Encountering Development: The Making and Unmaking of the Third World*, Princeton, NJ: Princeton University Press.

Esposito, R. (1998/2010) *Communitas. The Origin and Destiny of Community*, translated by T. Campbell, Stanford, CA: Stanford University Press.

Esposito, R. and Paparcone, A. (2006) 'Interview', *Diacritics*, 36.2, pp. 49–56.

Faulkner, J. (2001) 'Amnesia at the Beginning of Time: Irigaray's reading of Heidegger in The Forgetting of Air', *Contretemps* 2, May, pp. 124–141.

Feifer, M. (1985) *Going Places*, London: Macmillan.

Feld, S. (2005) 'Places Sensed, Senses Placed: Toward a Sensuous Epistemology of Environment', in David Howes (ed.), *The Sensual Culture Reader*, Oxford: Berg, pp. 179–191.

Fennell, D. (2006) *Tourism Ethics*, Clevedon: Channel View.

Forsyth, T. (2002) 'What Happened on "The Beach"? Social Movements and Governance of Tourism in Thailand', *International Journal of Sustainable Development*, 5.3, pp. 325–336.

Foucault, M. (2000) 'Different Spaces', in James Faubion (ed.), *Foucault, Aesthetics, Method, and Epistemology*, London: Penguin Books.

Foucault, M. (2000) 'Polemics, Politics and Problematizations', in P. Rabinow (ed.) *Essential Works of Foucault 1954–1984: Ethics, Subjectivity and Truth*, London: Penguin Books.

Foucault, M. (2000) 'The Masked Philosopher', in P. Rabinow (ed.) *Essential Works of Foucault 1954–1984: Ethics, Subjectivity and Truth*, London: Penguin Books.

Franklin, A. (2008) 'The Tourism Ordering. Taking Tourism More Seriously As a Globalising Ordering', *Civilisations*, LVII.1–2 – *Tourisme, mobilités et altérités contemporaines*, pp. 25–39.

Freire, P. (1998) *Pedagogy of Freedom: Ethics, Democracy, and Civic Courage*, Lanham, MD: Rowman & Littlefield.

Freire, P. (1970/2000) *Pedagogy of the Oppressed*, New York: Continuum.

Freire-Medeiros, B. (2009) 'The Favela and its Touristic Transits', *Geoforum*, 40.4, pp. 580–588.

Frenzel, F. (2011) 'Exit the System: Crafting the Place of Protest Camps Between Antagonism and Exception,' Working Paper. University of the West of England.

Friese, H. (2004) 'Spaces of Hospitality', *Journal of Theoretical Humanities*, 9.2, pp. 67–79.

Germann Molz, J. (2012) *Travel Connections: Tourism, Technology and Togetherness in a Mobile World*, London and New York: Routledge.

Germann Molz, J. and Gibson, S. (eds) (2007) *Mobilizing Hospitality. The Ethics of Social Relations in a Mobile World*, Aldershot: Ashgate.

Gibson, J. J. (1979) *The Ecological Approach to Visual Perception*, Boston, MA: Houghton Mifflin.

Gibson, S. (2003) 'Accommodating Strangers: British Hospitality and the Asylum Hotel Debate', *Journal for Cultural Research*, 7.4, pp. 367–386.

Girard, R. (1979) *Violence and the Sacred*, translated by P. Gregory, Baltimore, MD, and London: John Hopkins University Press.

Graburn, N. H. H. (1989) 'Tourism: The Sacred Journey', in Smith, V. L. (ed.), *Hosts and Guests: The Anthropology of Tourism*, 2nd edition, Philadelphia, PA: University of Pennsylvania Press, pp. 21–36.

Greene, S. (2003) 'Staged Cities: Mega-events, Slum Clearance, and Global Capital, *Yale Human Rights and Development Law Journal*, 6, pp. 161–179.

Grit, A. (2010) 'The Opening up of Hospitality Spaces to Difference: Exploring the Nature of Home Exchange Experiences', unpublished PhD thesis, University of Strathclyde.

Grit, A. and Lynch, P. (2011) 'Hotel Transvaal and Molar Lines As a Tool to Open up Spaces of Hospitality', in I. Ateljevic, N. Morgan and A. Pritchard (eds), *The Critical Turn in Tourism Studies: Creating an Academy of Hope*, London, Routledge, pp. 208–218.

Grit, A. and Lynch, P. (2012) 'An Analysis of the Development of Home Exchange Organisations', *Research in Hospitality Management*, 1.1, 2012, pp. 1–7.

Grit A. and Lynch, P. (2013) 'Moving towards Hospity', unpublished paper.

Grosz, E. (1995) 'Women, *Chora*, Dwelling', in *Space, Time and Perversion: Essays on the Politics of Bodies*, London and New York: Routledge.

Gunnison, E. (2012) 'How to Eat Ramen Like Anthony Bourdain', *Esquire*, http://www.esquire.com/blogs/food-for-men/anthony-bourdain-ramen-upgrades-8682344.

Hailey, C. (2008) *Campsite: Architectures of Duration and Place*, Baton Rouge, LA: Louisiana State University Press.

Hailey, C. (2009) *Camps: A Guide to 21st-Century Space*, Cambridge, MA: MIT Press.

Haraway, D. (1991) 'A Cyborg Manifesto: Science, Technology, and Socialist-Feminism in the 20th Century', in *Simians, Cyborgs and Women: The Reinvention of Nature*, New York: Routledge, pp. 149–181.

Hardt, M. and Negri, A. (2000) *Empire*, Cambridge, MA: Harvard University Press.

Harman, G. (2009) *The Prince of Networks: Bruno Latour and Metaphysics*, Melbourne: Re-Press.

Harrison P. (2007) 'The Space between Us: Opening Remarks on the Concept of Dwelling', *Environment and Planning D: Society and Space*, 25.4, pp. 625–647.

Harvey, D. (2005) *A Brief History of Neoliberalism*, New York: Oxford.

Heidegger, M. (1971/2001) *Poetry, Language, Thought*, translated by A. Hofstadter, New York: HarperCollins.

Heidegger, M. (1927/1972) *Sein und Zeit*. Tübingen: Max Niemeyer.

Heidegger, M. (1978/1993) 'Building Dwelling Thinking', in D. F. Krell (ed.), *Basic Writings: Revised and expanded edition*, London: Routledge, pp. 347–363.

Heikkilä, M. (2013) 'Monin vedoin: Nancy piirtämisen merkityksestä' [Multiple Traits: Jean-Luc Nancy on Drawing], *Tiede & Edistys*, 2, pp. 139–151.

Hiddleston, J. (2009) *Understanding Postcolonialism*, Stocksfield: Acumen Publishing Limited.

Higgins-Desbiolles, F. (2006) 'More Than an "Industry": The Forgotten Power of Tourism As a Social Force', *Tourism Management*, 27, pp. 1192–1208.

Höckert, E. (2011) 'Community-Based Tourism in Nicaragua: A Socio-Cultural Perspective', *The Finnish Journal of Tourism Research*, 7.2, pp. 7–25.

Hollinshead, K. (2002) 'Playing with the Past: Heritage Tourism Under the Tyrannies of Postmodern Discourse', in C. Ryan (ed.), *The Tourist Experience*, London: Continuum, pp. 172–200.

Horn, L. (1989) *A Natural History of Negation*, Chicago, IL: The University of Chicago Press.

Hultman, J. and Andersson Cederholm, E. (2010) 'Bed, Breakfast and Friendship: Intimacy and Distance in Small-Scale Hospitality Businesses', *Culture Unbound: Journal of Current Cultural Research*, 2, pp. 365–380.

Irigaray, L. (1994/2000) *Democracy Begins between Two*, London: The Athlone Press.

Iso-Ahola, S. E. (1982) 'Toward a Social Psychological Theory of Tourism Motivation: A Rejoinder', *Annals of Tourism Research*, 9.2, pp. 256–262.

Jokinen, E and Veijola, S. (1997) 'The Disoriented Tourist. The Figuration of the Tourist in Contemporary Cultural Critique', in C. Rojek and J. Urry (eds), *Touring Cultures: Transformations of Travel and Theory*, London and New York: Routledge, pp. 23–51.

Jokinen, E. and Veijola, S. (2003) 'Mountains and Landscapes: Towards Embodied Visualities', in D. Crouch and N. Lübbren (eds), *Visual Culture and Tourism*, Oxford: Berg, pp. 259–279.

Jokinen, E. and Veijola, S. (2012) 'Time to Hostess'. Reflections on Borderless Care', in C. Minca and T. Oakes (eds), *Real Tourism: Practice, Care and Politics in Contemporary Travel Culture*, London: Routledge, pp. 38–53.

Kapoor, I. (2004) 'Hyper Self Reflexive Development? Spivak on Representing the Third World "Other"', *Third World Quarterly*, 25.4, pp. 627–647.

Kearney, R. and Dooley, M. (eds) (1999) *Questioning Ethics*, London and New York: Routledge.

Kincaid, J. (1988) *A Small Place*, New York: Farrar, Strauss & Giroux.

Komter, A. and van Leer, M. (2012) 'Hospitality As a Gift Relationship: Political Refugees As Guests in the Private Sphere', *Hospitality & Society*, 2.1, pp. 7–23.

Krippendorf, J. (1987) *The Holiday Makers. Understanding the Impact of Leisure and Travel*, Oxford: Butterworth-Heinemann.

Kuokkanen, R. (2007) *Reshaping the University. Responsibilities, Indigenous Epistemes, and the Logic of the Gift*, Vancouver: UBC Press.

Kytö, M. (2013) Kotiin kuuluvaa. Yksityisen ja yhteisen kaupunkiäänitilan risteymiä. [Sounds like home: Crossings of private and common urban acoustic space]. Publications of the University of Eastern Finland. Dissertation in Education, Humanities and Theology. No: 45. Joensuu 2013.

Labelle, B. (2010) *Acoustic Territories: Sound Culture and Everyday Life*, London: Continuum.

Ladyga, Z. (2012) 'Tracing Levinas in Gayatri Spivak's Ethical Representations of Subalternity', in E. B. Luczak, J. Wierzchowska and J. Ziarkowska (eds), *In Other Words. Dialogizing Postcoloniality, Race, and Ethnicity*, Warzava: Encounters, pp. 221–230.

Larsen, J., Urry, J. and Axhausen, K. W. (2007) 'Networks and Tourism: Mobile Social Life', *Annals of Tourism Research*, 34.1, pp. 244–262.

Lashley, C. and Morrison, A. (eds) (2000) *In Search of Hospitality: Theoretical Perspectives and Debates*, Oxford: Butterworth-Heinemann.

Latimer, J. (2013) 'Being Alongside. Rethinking Relations amongst Different Kinds', *Theory, Culture & Society*, 30.7–8, pp. 77–104.

Latour, B. (2013) *An Inquiry into Modes of Existence: An Antropology of the Moderns*, translated by C. Porter, Cambridge, MA: Harvard University Press.

Law, L., Bunnell, T. and Ong, C.-E. (2007) '*The Beach*, Gaze and Film Tourism', *Tourist Studies*, 7.2, pp. 141–164.

Lenz, R. (2010) '"Hotel Royal" and Other Spaces of Hospitality: Tourists and Migrants in the Mediterranean', in J. Scott and T. Selwyn (eds), *Thinking Through Tourism*, Oxford: Berg, pp. 209–230.

Lerup, L. (1977) *Building the Unfinished: Architecture and Human Action*, London: Sage.

Levinas, E. (1969) *Totality and Infinity: An Essay on Exteriority*, translated by A. Lingis, Pittsburgh, PA: Duquesne University Press.

Levinas, E. (1985) *Ethics and Infinity: Conversations with Philippe Nemo*, translated by R.A. Cohen. Pittsburgh, PA: Duquesne University Press.

Levinas, E. (1987) *Time and the Other and Additional Essays*, translated by R.A. Cohen, Pittsburgh, PA: Duquesne University Press.

Levinas, E. (1998) *Otherwise than Being, or Beyond Essence*, translated by A. Lingis, Pittsburgh, PA: Duquesne University Press.

Lewis, T. and Lortie, M.-C. (2013) 'Cook It Raw: For chefs, it's like free-falling into the unknown', *The Observer*, Saturday 20 April 2013, http://www.guardian.co.uk/lifeandstyle/2013/apr/21/cook-it-raw-eight-chefs-recall.

Lingis, A. (1994) *The Community of Those Who Have Nothing in Common*, Bloomington and Indianapolis, IN: Indiana University Press.

Löfgren, O. (1999) *On Holiday: A History of Vacationing*, Berkeley, CA: University of California Press.

Lynch, P. (2005) 'Sociological Impressionism in a Hospitality Context', *Annals of Tourism Research*, 32.3, pp. 527–548.

Lynch, P., Germann Molz, J., McIntosh, A., Lugosi, P. and Lashley, C. (2011) 'Theorising Hospitality', *Hospitality & Society*, 1.1, pp. 3–24.

Määttänen, K. (1998) 'Sense of Self and Narrated Mothers in Women's Autobiographies', in Satu Apo, Aili Nenola and Laura Stark-Arola (eds), *Gender and Folklore: Perspectives on Finnish and Karelian Culture*, Studia Fennica Folkloristica 4, Helsinki: Finnish Literature Society, pp. 317–331.

MacCannell, D. (1976/1999) *The Tourist: A New Theory of the Leisure Class*, Berkeley, CA: University of California Press.

Malins, P. (2004) 'Machinic Assemblages: Deleuze, Guattari and an Ethico-Aesthetics of Drug Use', *Janus Head*, 7.1, pp. 84–104.

Marcus, G. and Clifford, J. (eds) (1986) *Writing Culture: The Poetics and Politics of Ethnography*, London: University of California Press.

Marshall, J. P. and Goodman, J. (2013) 'Disordering Network Theory: An Introduction', *Global Networks*, 13.3, pp. 279–289.

Mazzullo, N. and Ingold, T. (2008) 'Being Along: Place, Time and Movement among Sámi People', in J. O. Baerenholdt and B. Granås (eds), *Mobility and Place: Enacting Northern European Peripheries*, Aldershot: Ashgate, pp. 27–38.

McEwan, C. (2009) *Postcolonialism and Development*, New York: Routledge.

Medovoi, L., Raman, R. and Robinson, B. (1990) 'Can the Subaltern Vote?', *Socialist Review*, 20.3, pp. 133–149.

Metcalfe, A. and Game, A. (2008) 'Potential Space and Love', *Emotion, Space and Society* 1, pp. 18–21.

Minca, C. (2010) 'The Island: Work, Tourism and the Biopolitical', *Tourist Studies*, 9.2, pp. 88–108.

Minca, C. (2011) 'No Country for Old Men', in C. Minca and T. Oakes (eds), *Real Tourism: Practice, Care and Politics in Contemporary Travel Culture*, London: Routledge, pp. 2–37.

Moraña, M., Dussel, E. and Jáuregui, C. A. (eds) (2008) *Coloniality at Large: Latin America and the Postcolonial Debate*, Durham: Duke University Press.

More, G. (2011) *The Politics of the Gift: Exchanges in Poststructuralism*, Edinburgh: Edinburgh University Press.

Morgan, N. and Cole, S. (eds) (2010) *Tourism and Inequality: Problems and Prospects*, Oxfordshire: CABI.

Morton, S. (2007) *Gayatri Spivak. Ethics, Subalternity and the Critique of the Postcolonial Reason*, Cambridge: Polity Press.

Nancy, J.-L. (1990) *La communauté désoeuvrée*, 2nd edition, Paris: Bourgois.

Nancy, J.-L. (2000) *Being Singular Plural*, translated by R. D. Richardson and A. E. O'Byrne, Stanford, CA: Stanford University Press.

Nancy, J.-L. (2009) *Le Plaisir au dessin*, Paris: Galilée.

Nancy, J.-L. (1992/1995) *Corpus*, translated by Susanna Lindberg, Tampere: Gaudeamus.

Nygren, A. (1999) 'Local Knowledge in the Environment-Development Discourse. From Dichotomies to Situated Knowledges', *Critique of Antropology*, 19.3, pp. 267–288.

O'Gorman, K. (2007) 'Discovering Commercial Hospitality in Ancient Rome', *Hospitality Review*, 9.2, pp. 44–52.

O'Gorman, K. and Lynch, P. A. (2008) 'Monastic Hospitality: Explorations', University of Strathclyde Institutional Repository, 2008, http://core.kmi.open.ac.uk/display/9020089.

Pantzar, M. (2010) 'Future Shock – Discourses Changing Temporal Architecture of Daily Life', *Journal of Future Studies*, 14.4, pp. 1–22.

Pérez, F. P., Barrera. O. P., Peláez, A. V. and Lorío, G. (2010) *Turismo Rural Comunitario como Alternativa de reducción de la pobreza rural en Centroamérica*. Managua: EDITASA.

Pine, B. J. and Gilmore, J. H. (1999) *The Experience Economy: Work Is Theatre & Every Business a Stage*, Boston, MA: Harvard Business School Press.

'Plan Nacional de Desarrollo. Bolivia Digna, Soberana, Productiva y Democrática para VivirBien. Lineamientos Estratégicos 2006–2011. República de Bolivia. Ministerio de Planificación del Desarrollo, La Paz, Bolivia, Septiembre, 2007.'

Pløger, J. (2006) 'In Search of Urban Vitalis', *Space and Culture*, 9.4, pp.382–399.

Pozorov, S. (2010) 'Why Giorgio Agamben is an Optimist', *Philosophy Social Criticism*, 36, pp. 1053–1073.

Pratt, M. L. (1992) *Imperial Eyes Travel Writing and Transculturation*, London and New York: Routledge.

Pritchard, A., Morgan, N. and Ateljevich, I. (2011) 'Hopeful Tourism, A New Transformative Perspective', *Annals of Tourism Research*, 38.3, pp. 941–963.

Puumala, E. (2013) 'Politiikan tuntu, mieli ja merkitys. Tapahtuva yhteisö ja poliittisen kokemus kehollisissa kohtaamisissa' [The Senses of Politics: Corporeal Junctures and the Experience of the Political], *Tiede & Edistys*, 2, pp. 125–138.

Pyyhtinen, O. (2009) 'Being-with: Georg Simmel's Sociology of Association', *Theory, Culture & Society*, 26.108, pp. 108–128.

Pyyhtinen, O. (2010) *Simmel and 'the Social'*, Basingstoke and New York: Palgrave Macmillan.

Pyyhtinen, O. (2014) *The Gift and its Paradoxes. Beyond Mauss*, Surrey: Ashgate.

Pyyhtinen, O. and Lehtonen, T.-K. (2014) 'Michel Serres ja yhteisön logiikat [Michel Serres and the Logics of Community]', in I. Kauppinen and M. Pyykkönen (eds), 1900-luvun ranskalainen yhteiskuntateoria, Gaudeamus: Helsinki. (To be published.)

Raffoul, F. (1998) 'The Subject of the Welcome. On Jacques Derrida's Adíeu á Emmanuel Levinas', *Symposium*, 2.2, pp. 211–222.

Rantala, O. and Valtonen, A. (2014) 'A Rhythmanalysis of Touristic Sleep in Nature', *Annals of Tourism Research*, 47, pp. 18–30.

Ranta-Owusu, E. (2010) 'Governing Pluralities in the Making. Indigenous Knowledge and the Question of Sovereignty in Contemporary Bolivia', *Journal of the Finnish Anthropological Society*, 35.3, pp. 28–48.

Read, J. (2009) 'A Genealogy of Homo Economicus: Neoliberalism and the Production of Subjectivity', *Foucault Studies*, 6, pp. 25–36.

Romero, S. (2012) 'Slum Dwellers Are Defying Brazil's Grand Design for Olympics', *The New York Times*, 4 March 2012.

Rumford, C. (2013) *The Globalization of Strangeness*, Chippenham and Eastbourne: Palgrave Macmillan.

Saarinen, J. (2006) 'Traditions of Sustainability in Tourism Studies', *Annals of Tourism Research*, 33.4, pp. 1121–1140.

Searle, J. (1969/1996) *Speech Acts. An Essay in the Philosophy of Language*. Cambridge: Cambridge University Press.

Serres, M. (1982) 'Platonic dialogue', in J. V. Harari and D. F. Bell (eds), *Hermes: Literature, Science, Philosophy*, Baltimore, MD, and London: John Hopkins University Press, pp. 65–70.

Serres, M. (1993/1995) *Angels, A Modern Myth*, translated by F. Cowper, Paris: Flammarion.

Serres, M. (1995) *Genesis*, translated by G. James and J. Nielson, Ann Arbor, MI: The University of Michigan Press.

Serres, M. (1995) *The Natural Contract*, translated by E. MacArthur and W. Paulson, Ann Arbor, MI: The University of Michigan Press.

Serres, M. (2007) *The Parasite*, translated by L. R. Schehr, Minneapolis, MN: University of Minnesota Press.

Serres, M. (2011) *Malfeasance. Appropriation through Pollution?* translated by A.-M. Feenberg-Dibon, Stanford, CA: Stanford University Press.

Serres, M. (2014) *Times of Crisis: What The Financial Crisis Revealed and How to Reinvent our Lives and Future*, translated by A.-M. Feenberg-Dibon, New York and London: Bloomsbury.

Serres, M. with Latour B. (1995) *Conversations on Science, Culture, and Time*, translated by R. Lapidus, Michigan: University of Michigan Press.

Sharpe, J. and Spivak, G. C. (2003) 'A Conversation with Gayatri Chakravorty Spivak: Politics and the Imagination', *Signs*, 28.2, pp. 609–624.

Sharpley, R. (2011) *Past Trends and Future Directions*, Oxon: Routledge.

Sharr, A. (2007) *Heidegger for Architects*, London and New York: Routledge.

Shaviro, S. (2009) *Without Criteria: Kant, Whitehead, Deleuze, and Aesthetics*, Cambridge, MA: MIT Press.

Shaviro, S. (2010) *Post-Cinematic Affect*, Winchester and Washington: Zero books.

Simmel, G. (1902/1903) 'The Number of Members As Determining the Sociological Form of the Group', *The American Journal of Sociology*, VIII.I and I, pp. 1–46, 158–196.

Simmel, G. (1908/1992) *Soziologie*, Georg Simmel Gesamtausgabe, Band 11, Frankfurt am Main: Suhrkamp.

Simmel, G. (1971) 'The Metropolis of Modern Life', in D. Levine (ed.), *Simmel: On Individuality and Social Forms*, Chicago, IL: Chicago University Press.

Simmel, G. (1999) 'Grundfragen der Soziologie', in G. Fitzi and O. Rammstedt (eds), *Georg Simmel Gesamtausgabe*, Vol. 16, Frankfurt am Main: Suhrkamp, pp. 59–179.

Simmel, G. (1917/1920) 'Die Geselligkeit', in *Grundfragen der Soziologie*, Leipzig: Sammlung Göschen.

Smith, D. W. (2007) 'Deleuze and the Question of Desire: Toward an Immanent Theory of Ethics', *Parrhesia*, 2/2007, pp. 66–78.

Smith, M. (2009) 'Ethical Perspectives: Exploring the Ethical Landscape of Tourism', in T. Jamal and M. Robinson (eds), *The SAGE Handbook of Tourism Studies*, London: Sage, pp. 613–630.

Smith V. L. (ed.) (1989) *Hosts and Guests: The Anthropology of Tourism*, Philadelphia, PN: University of Pennsylvania Press.

Spivak, G. C. (1987) *In Other Worlds: Essays in Cultural Politics*, London: Routledge.

Spivak, G. C. (1988) 'Can the Subaltern Speak?', in C. Nelson and L. Grossberg (eds), *Marxism and Interpretation of Culture*, Chicago, IL: University of Illinois Press, pp. 271–313.

Spivak, G. C. (1993) *Outside in the Teaching Machine*, New York: Routledge.

Spivak, G. C. (1995) 'Translator's Preface and Afterword', in *Imaginary Maps: Three Stories*, London: Routledge.

Spivak, G. C. (1999) *A Critique of Postcolonial Reason: Toward a History of the Vanishing Present*, Cambridge, MA: Harvard University Press.

Spivak, G. C. (2001) 'A Note on the New International', *Parallax*, 7.3, pp. 12–16.

Still, J. (2006) 'France and the Paradigm of Hospitality', *Third Text*, 20. 6, pp. 703–710.

Strhan, A. (2012) *Levinas, Subjectivity, Education. Towards an Ethics of Radical Responsibility*, West Sussex: Wiley Blackwell.

Taylor, J. (2001) 'Authenticity and Sincerity in Tourism', *Annals of Tourism Research*, 28.1, pp. 7–26.

Teivainen, T. (2004) *Pedagogía del Poder Mundial*, INTO eBooks.

The Wordsworth Thesaurus (1993) Denmark: Wordsworth Reference.

Till, J. (2009) *Architecture Depends*, Cambridge, MA: MIT Press.

Tolkien, J. R. R. (1954) *The Lord of the Rings*, London: Harper Collins.

Truax, Barry (2011) *Acoustic Communication*, 2nd edition, Wesport: Ablex.

Turner, V. (1969/1977) *The Ritual Process. Structure and Anti-structure*, Ithaca, NY: Cornell University Press.

Turner, V. (1982) *From Ritual to Theatre. The Human Seriousness of Play*, New York: PAJ Publications.

Tzanelli, R. (2006) 'Reel Western Fantasies: Portrait of a Tourist Imagination in *The Beach* (2000)', *Mobilities*, 1.1, pp. 121–142.

Tzanelli, R. (2007) *The Cinematic Tourist: Explorations in Globalization, Culture and Resistance*, London and New York: Routledge.

United Nations World Tourism Organization (UNWTO) (2004) *Making Tourism More Sustainable: A Guide for Policy Makers*, Madrid: United Nations World Tourism Organization.

Urry, J. (1990) *The Tourist Gaze*. London: Sage.

Valtonen, A. and Veijola, S. (2011) 'Sleep in Tourism', *Annals of Tourism Research*, 38.1, pp. 175–192.

Veijola, S. (1994) 'Metaphors of Mixed Team Play', *International Review for the Sociology of Sport*, 29.1, pp. 32–49.

Veijola, S. (1997) 'Luku, suku ja sosiaalinen: Taipuuko varsinainen sosiaalinen myös naissuvun mukaan?' [Number, Gender and the Social: Can the Proper Social Be Conjugated Even in Feminine Gender?], *Naistutkimus/ Kvinnoforskning*, 4, pp. 2–29.

Veijola, S. (2004) 'Pelaajan ruumis. Sekapeli modaalisena sopimuksena'[The Body of the Player. Mixed Team Play as a Modal Contract], in Eeva Jokinen, Marja Kaskisaari and Marita Husso (eds), *Ruumis töihin! Käsite ja käytäntö* [The Body to Work! Concept and Practice], Tampere: Vastapaino, pp. 99–124.

Veijola, S. (2005) 'Turistien yhteisöt' [Tourist Communities], in Hautamäki, A., Lehtonen, T., Sihvola, J., Tuomi, I., Vaaranen, H. and Veijola, S. (eds) *Yhteisöllisyyden paluu* [The Return of Communality], Helsinki: Gaudeamus, pp. 90–113.

Veijola, S. (2006) 'Heimat Tourism in the Countryside. Paradoxical Sojourns in Self and Place', in T. Oakes and C. Minca (eds), *Travels in Paradox: Remapping Tourism*. New York: Bowman & Littlefield, pp. 77–95.

Veijola, S. (2009) 'Gender As Work in the Tourism Industry', *Tourist Studies*, 9.2, pp. 109–126.

Veijola, S. and Falin, P. (2014) 'Mobile Neighbouring', *Mobilities*, online first, doi 10.1080/17450101.2014.936715.

Veijola, S. and Jokinen, E. (1994) 'The Body in Tourism', *Theory, Culture & Society*, 11.3, 125–151.

Veijola, S. and Jokinen, E. (1998) 'The Death of the Tourist', *European Journal of Cultural Studies*, 1.3, pp. 327–351.

Veijola, S. and Jokinen, E. (2008) 'Towards a Hostessing Society? Mobile Arrangements of Gender and Labour', *NORA: Nordic Journal of Feminist and Gender Research*, 16.3, pp. 166–181.

Vikman, N. (2007) *Eletty ääniympäristö. Pohjoisitalialaisen Cembran kylän kuulokulmat muutoksessa* [*The lived acoustic environment. Cembra's changing points of ear*]. Tampere: Acta Universitatis Tamperensis, 1271.

Virilio, P. (2010) *The Futurism of the Instant. Stop-Eject*, Cambridge: Polity Press.

von Wright, G. H. (1977) 'Deonttinen logiikka' [Deontic Logics], in Tauno Nyberg (ed.), *Ajatus ja analyysi*, WSOY: Helsinki, pp. 147–166.

Vrasti, W. (2013) *Volunteer Tourism in the Global South: Giving Back in Neoliberal Times*, London and New York: Routledge.

Wearing, B. (1998) *Leisure and Feminist Theory*, London: Sage.

Weaver, D., Buckley, R., Wheeller, B. and Bramwell, B. (2012) 'Mass Tourism and Sustainability: Can the Two Meet?, in Tej Vir Singh (ed.), *Critical Debates in Tourism*, Bristol: Channel View, pp. 27–52.

White, N. R. and White, P. B. (2008) 'Travel As Interaction: Encountering Place and Others', *Journal of Hospitality and Tourism Management*, 15.3, pp. 42–48.

Whitehead, A. N. (1929/1978) *Process and Reality*. Corrected edition. New York: Free Press.

Whyte, J. (2010) 'A New Use of the Self: Giorgio Agamben on the Coming Community', *Theory and Event*, 13.1, pp. 1–19.

Williams, J. (2013) *Gilles Deleuze's Difference and Repetition: A Critical Introduction and Guide*, Edinburgh: Edinburgh University Press.

Winnicot, D. W. (1991) *Play and Reality*, London: Routledge.

Wittel, A. (2001) 'Toward a Network Sociality', *Theory, Culture & Society*, 18.6, pp. 51–76.

World Commission on Environment and Development (WCED) (1987) *Our Common Future*, Oxford: Oxford University Press.

Yates, J. (2005) '"The Gift is a Given". On the Errant Ethic of Michel Serres', in N. Abbas (ed.), *Mapping Michel Serres*, Ann Arbor, MI: The University of Michigan Press, pp. 190–209.

Young, I. M. (1980) 'Throwing Like a Girl. A Phenomenology of Feminine Body Comportment, Motility and Spatiality', *Human Studies*, 3.2, pp. 137–156.

Zapata, M. J., Hall, C. M., Lindo, P. and Vanderschaeghe, M. (2012) 'Can Community-Based Tourism Contribute to Development and Poverty Alleviation? Lessons from Nicaragua', in J. Saarinen, C. M. Rogerson and H. Manwa (eds), *Tourism and the Millennium Development Goals: Tourism, Local Communities and Development*, London: Routledge, pp. 98–122.

Zimmerman, M. E. (1993) 'Heidegger, Buddhism and Deep Ecology', in C. B. Guignon (ed.), *The Cambridge Companion to Heidegger*, Cambridge: Cambridge University Press.

Index

Printed and bound in the United States of America